Gerry Stahl's assembled texts volume #6

Constructing Dynamic Triangles Together (pre-publication version)

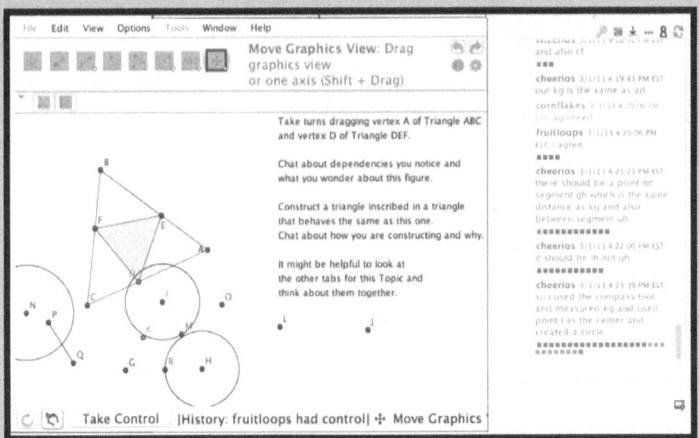

Gerry Stahl

Gerry Stahl's Assembled Texts

1. *Marx and Heidegger*

2. *Tacit and Explicit Understanding in Computer Support*

3. *Group Cognition: Computer Support for Building Collaborative Knowledge*

4. *Studying Virtual Math Teams*

5. *Translating Euclid: Designing a Human-Centered Mathematics.*

6. *Constructing Dynamic Triangles Together: The Development of Mathematical Group Cognition*

7. *Essays in Social Philosophy*

8. *Essays in Personalizable Software*

9. *Essays in Computer-Supported Collaborative Learning*

10. *Essays in Group-Cognitive Science*

11. *Essays in Philosophy of Group Cognition*

12. *Essays in Online Mathematics Interaction*

13. *Essays in Collaborative Dynamic Geometry*

14. *Adventures in Dynamic Geometry*

15. *Global Introduction to CSCL*

16. *Editorial Introductions to ijCSCL*

17. *Proposals for Research*

18. *Overview and Autobiographical Essays*

19. *Theoretical Investigations*

20. *Works of 3-D Form*

21. *Dynamic Geometry Game for Pods*

Gerry Stahl's assembled texts volume #6

Constructing Dynamic Triangles Together (pre_publication version)

Gerry Stahl

Gerry Stahl
Gerry@GerryStahl.net
www.GerryStahl.net

ISBN 978-1-105-38981-8 (paperback)
ISBN 978-1-105-62548-0 (ebook)

Pre-publication version

This volume is a pre-publication version of *Constructing Dynamic Triangles Together: The Development of Mathematical Group Cognition*, published by Cambridge University Press in 2015. These materials were last revised March 16, 2015, from the final manuscript. This version has not been edited, laid out or paginated by Cambridge University Press. Please do not cite page numbers from this version or quote from it. This version is only for informal use and may not be duplicated. Please refer to the Cambridge University Press version for official usage, citation and pagination.

Errata of the published book

These are the only errata known to the author as of the latest revision of this pre-publication version. They have been corrected in this version. Please notify the author at Gerry@GerryStahl.net if you discover any additional errors.

●

Introduction

Rational thinking as exemplified in mathematical cognition is of undeniable importance in the modern world. This book documents how a group of three eighth-grade girls developed specific practices typical of such thinking through involvement in an online educational experience. The presentation begins by discussing the methodological approach adopted in analyzing the development of mathematical group cognition. An extended case study then tracks the team of students step by step through its eight-hour-long progression. Concluding sections draw the consequences for the theory of group cognition and for educational practice.

The book investigates the display of mathematical reasoning by the students discussing dependencies within a sequence of dynamic-geometry figures. By examining the network of their mutual chat responses preserved in computer logs coordinated with their geometric actions exhibited in a replayer, it is possible to follow in detail the meaning-making processes of the students and to observe how the team develops its mathematical group cognition by adopting a variety of group practices. The longitudinal data set provides a rich opportunity to observe cognitive development through the interplay of processes and practices identifiable at the individual, small-group and community units of analysis.

The examination of data focuses on these areas of the team's development:

i.   Its effective team *collaboration*,

ii.   Its productive mathematical *discourse*,

iii. Its enacted use of dynamic-geometry *tools* and

iv.   Its ability to identify and construct dynamic-geometry dependencies by:

      a.   Dynamic *dragging* of geometric objects,

      b.   Dynamic *construction* of geometric figures and

      c.   Dynamic design of *dependencies* in geometric relationships.

The analysis reveals how the three students contribute differently, but also appropriate each other's contributions. This successively advances the group's ability to collaborate effectively with group agency, to articulate mathematical ideas productively by applying increasingly meaningful mathematical terminology and to engage in dynamic-geometry challenges using mastered software functionality. The shared digital workspace supports group exploration and testing of geometric conjectures, while sequenced curricular topics guide student discoveries. These affordances help the students to advance to new levels of individual and group mathematical cognition through the situated adoption of many specific group practices for productive collaboration, mathematical discourse and dynamic-geometry problem solving.

The result is a detailed case study of the Virtual Math Teams Project as a paradigmatic example of computer-supported collaborative learning, incorporating a unique model of human-computer interaction analysis applied to the use of innovative educational technology.

— Philadelphia, March 16, 2015

Contents

Introduction to the Analysis

Designing computer support for the learning of mathematics is a major educational challenge today. Networked computers provide an attractive opportunity to explore collaborative-learning approaches to math education. The recent availability of dynamic-geometry software provides further opportunity for innovation. This book reports on an extensive research effort involving teaching math teachers and their students in an online collaboration environment. Specifically, it documents the cognitive development of a particular team of three students learning about dynamic geometry in that virtual social setting. An extended case study shows how the team enacts software tools and adopts group practices within the educational research project, which was designed to extend and support their ability to collaborate, to engage in mathematical discourse and to explore or construct dynamic-geometric figures. The book provides detailed empirical support, within a math-education context, for the theory and practice of group cognition.

Research Context

This volume builds on earlier publications about the Virtual Math Teams (VMT) Project, putting their arguments into practice, documenting their claims, fleshing out their theory and fulfilling their promises. It culminates a cycle of books reporting on the project:

- *Group Cognition* (Stahl, 2006, MIT Press) introduced the VMT Project as a response to practical and foundational issues in CSCL (computer-supported collaborative learning) and CSCW (computer-supported cooperative work). It recommended adapting methods of interaction analysis to online text chat. It proposed that the investigation of small-group processes and practices could provide insight into online collaborative learning. It outlined a preliminary theory of group cognition as a framework appropriate to computer-mediated interaction.

- *Studying Virtual Math Teams* (Stahl, 2009, Springer) described the technology approach and affordances of the VMT software environment. The edited volume provided illustrative analyses of brief excerpts of student interaction in VMT by a number of international researchers. It suggested technology-design features and methodological considerations. It expanded the philosophic and scientific basis of group-cognition theory.

- *Translating Euclid* (Stahl, 2013, Morgan & Claypool) reviewed the VMT Project: its multi-user technology, collaborative pedagogy, dynamic-geometry curriculum, design-based research approach and educational goals. The multi-faceted research project was situated within its historical, mathematical and educational context. This recent project review discussed the integration of collaborative dynamic geometry into the VMT environment. It further elaborated the theory of group cognition as a basis for educational innovation.

The present book documents the findings of the VMT Project as a paradigmatic example of Computer-Supported Collaborative Learning (CSCL) exploration, incorporating a unique model of Human-Computer Interaction (HCI) analysis. Directed by the author for the past twelve years, the VMT Project pioneered a method of analyzing interaction data, adapting ethnomethodologically inspired interaction analysis to the special conditions of computer-mediated collaboration and to the needs of design-based research in mathematics education. This fine-grained report on data from the VMT Project applies its methods longitudinally to the full eight hours of one student group's interaction. In this analysis, it details the team's cognitive development. It ties the development of their group cognition to the technological mediation, which takes place at multiple levels of the project:

- The students interact exclusively through the VMT online collaboration environment using text chat.
- The student team explores dynamic geometry in a computer simulation.
- The domain of dynamic geometry is defined by its software implementation.
- The VMT curriculum is technologically scripted for use with minimal teacher intervention.
- All the data is collected electronically through comprehensive instrumentation of the collaboration environment.

The centrality of computer support to the project makes this book relevant to (i) CSCL, (ii) HCI, (iii) mathematics instruction and (iv) educational technology:

(i)   From a CSCL perspective, this book is paradigmatic in offering a detailed example of research based on the theory of group cognition. The cognitive development of the observed team of students is conceived as computer-supported collaborative learning, in which learning is primarily viewed at the small-group unit of analysis of collaboration and all the communication takes place through computer-mediated interaction. It provides a rich picture of learning on many levels, not just measuring a single learning outcome. It not only documents *that* learning took place by the student team, but also details *how* the learning happened by observing the enactment of numerous group practices. It provides an examination of small-group cognitive development in terms of the adoption of group practices, including the enactment of tools. This approach is framed in the philosophy of group cognition, which has emerged from the VMT Project and is grounded in its findings. A rich picture of a prototypical instance of computer-supported collaborative learning emerges from this research.

(ii)   From an HCI standpoint, the book's analysis is distinctive in that it documents an investigation in which computer-mediated interaction analysis played a central role in the design-based research process, providing feedback to the project at multiple points: advice to teachers between sessions, revisions for the next cycle and formative evaluation of the overall project, including elaboration of the theoretical framework.

(iii)   From a mathematics instruction view, the book offers several proposals. In terms of curriculum design, the set of topics illustrates a focus on a central theoretical concept of the domain: dependency relationships in dynamic geometry. The online presentation of the topics to small groups of students illustrates a form of guidance toward mathematical understandings through computer scripting or scaffolding, with minimal direct teacher intervention. The sequential accumulation of group practices provides a conceptualization of increasing mathematical understanding. Finally, the collaborative approach to work on challenging problems reveals the mutual contributions from student zones of proximal development, which are negotiated and adopted by the group as cognitive practices.

(iv)   From an educational-technology approach, the book is unique in offering a longitudinal case study, which details cognitive development starting when the students first encounter online collaborative dynamic geometry. It identifies dozens of group practices by which the team of students learns to collaborate, to enact software tools, to understand geometric figures and to discuss mathematical invariants and their dependencies. It thereby shows how an online collaboration environment can facilitate learning—specifically the critical development of geometric reasoning—by providing a supportive space for the emergence and adoption of group practices.

As the concluding volume reporting on the VMT Project, this book illustrates a successful implementation of group-cognition research and analysis. Since it was proposed in the 2006 volume, the theory of group cognition has been increasingly accepted within the research community as an alternative to the traditional educational-psychology approach to instructional technology, focused on measurable learning outcomes of individual minds. As a presentation of CSCL methodology, the book provides an alternative or complement to statistical coding approaches. Within HCI, it shows that an ethnomethodologically informed approach can generate implications for design systematically within a practical design-based research process. Within the mathematics-instruction literature, it offers several proposals concerning curriculum focus on underlying relationships, guidance toward mathematical principles, operationalizing deep understanding in terms of practices and appreciating mechanisms of collaborative learning of mathematics. As an educational-technology intervention, it demonstrates the potential and details the challenges of using collaborative dynamic-geometry software to facilitate the development of mathematical cognition.

Presentation Structure

Constructing Dynamic Triangles Together: The Development of Mathematical Group Cognition rounds out the story of the VMT Project. It centers on an extended case study: the detailed longitudinal analysis of eight hours of interaction by a virtual math team of three middle-school girls working on an introductory sequence of dynamic-geometry challenges. It fulfills the promises and claims of previous publications on VMT by demonstrating the success of the methods they proposed, and carrying out systematic analysis of one team's entire online collaborative-learning experience. Along the way, it provides lessons for online curricular design, for CSCL technology and for HCI analysis. It also fills in the theory of group cognition with concrete results based on detailed data showing how collaborative learning of mathematics takes place through the enactment of specific group practices for collaboration, math discourse and software tool usage.

Attempts to study collaborative learning are often confounded by ambiguity about what the learners already know. Even more generally, evidence of various factors affecting the learning are missing from the available data. For instance, there may be social influences or power relationships that are not captured in the data or there may have been interactions, gestures and speech that were off-camera or unintelligible. Even worse, self-reports and introspection about learning take place long after foundational instances of learning have been processed, transformed and internalized. The learning analyzed in this book, in contrast, involves the students' initial encounters with a subject that is new for them: geometry, especially dynamic-geometry construction. Furthermore, their interactions about these encounters are captured live in full logs and replayer files, which reproduce the interactions just as they were present to the students. We assume that the students had previous familiarity with the visual appearances of conventional basic shapes of everyday geometry, but we are interested in how the team develops beyond this knowledge. There are certainly other influences on the individual mental activity of the students, based on their past and on events not captured in the VMT system, but we are focused on the team's development at the group unit of analysis; everything that took place between the students and was shared by the team passed through the VMT technology and was logged. So the data analyzed here is about as complete as one could hope for and as required by our methodology. To the extent practical, the VMT data documents the beginnings of mathematical cognition in the domain of introductory dynamic geometry for the team.

The team's developmental trajectory during their VMT experience is guided by a carefully designed sequence of curricular units: the topics that the students worked on in their eight sessions. The following analysis considers the team's work on each topic in order. The topics are planned to introduce the students methodically to the fundamentals of dynamic geometry. In particular, the goal is to have the team develop an understanding of dependency relationships that establish invariances, such as the maintained equality of side lengths of an equilateral triangle. The curriculum builds systematically. It starts by letting the students play with the most basic steps of construction, while guiding the team to work collaboratively. It introduces the building of an equilateral triangle as a prototypical construction and then extends it for the construction of perpendicular bisectors and right triangles. Because an understanding of problems and solutions in dynamic geometry is mediated by ones mastery of the software tools for manipulating and constructing dynamic-geometry objects, the most important tools are introduced before the topics that require them. As the team explores the use of the tools and engages in problem solving in response to the curricular topics, the team starts to adopt group practices. The analysis of the team interaction focuses on how the team enacts the tools and it identifies various kinds of practices that the team adopts.

The adopted group practices are taken to be important constituents of the team's group cognition. The team learns by successively embracing specific practices. For instance, in its early sessions, the students learn to work together effectively by incorporating group collaboration practices. These practices are in part suggested by the curriculum. The team negotiates them and then begins to follow them. Similarly, they gradually integrate group mathematical practices—often involving using the software tools to drag and construct dynamic-geometry figures—into their joint work. These practices establish necessary foundations for computer-supported collaborative learning in this domain of mathematics.

By identifying the team's adoption of group practices, the analysis in this book provides a paradigmatic example of CSCL. The case study analyzes the computer technology, as enacted by the team. It shows the mediation of the team's interaction by the integrated online pedagogy and domain-centered curriculum. It focuses on interaction at the group unit of analysis, and illustrates the methodological approach of the theory of group cognition. Its longitudinal approach provides a rich example of how collaborative learning can take place, while suggesting design lessons for improving the next iteration of software, pedagogy, curriculum, analysis and theory.

Overall, the detailed and extended longitudinal case study provides a rare view into how students learn in small groups. The many individual actions described are united into a narrative about the development of mathematical group cognition, framed in a theoretical and methodological perspective and leading to pedagogical and curricular lessons.

The presentation is divided into a number of chapters. The bulk of the volume conducts a fine-grained analysis of the student interaction and identifies the team's adoption of group practices into their interaction. These analysis chapters illustrate many aspects of sequential-interaction analysis, show how the students enact the use of the available technology, examine the student interpretation of curricular artifacts and display the student engagement in specific group practices. Each analysis chapter concludes with an assessment of the team's cognitive development and a set of implications for redesign of project details. This core of the book is preceded by methodological considerations and followed by theoretical reflections.

The chapters are:

- *Researching Mathematical Cognition.* The initial methodological chapter emphasizes the importance of mathematical cognition in the modern world and the difficulty it presents for many students. It briefly considers issues of schooling and theories concerning the development of mathematical understanding. It then argues for a case-study approach, incorporating sequential-interaction analysis. Building on Vygotsky's ideas, it suggests focusing on developmental processes at the group unit of analysis.

- *Analyzing Development of Group Cognition.* The VMT Project is described in the following chapter as design-based research, which incorporates cycles of refining technology, curriculum and theory through iterative trials with classroom teachers and students. The goals of the project—providing the focus of analysis in this book—include: development of collaboration skills, mathematical discourse and usage of software tools. Dynamic geometry is briefly described, with its characteristics of dragging, constructing and defining mathematical dependencies. The analytic methodology is then presented as sequential-interaction analysis, with a special emphasis on extended sequences of interaction involved in geometric problem solving. Such analysis can highlight the display by students of their collaborative mathematical development as they chat, manipulate graphical objects, explore problems, construct geometric figures and articulate solutions. In this way, analysis not only indicates that certain learning transpired, but also shows how it took place: through the adoption of group practices.

- Session *1: The Team Develops Collaboration Practices.* This first analysis chapter shows how the three students developed into a collaborative team, largely during their initial hour together online. At first, the students had no idea what to do in the VMT environment. However, they successively responded to suggestions within the environment—textual instructions, software displays, results of explorations. The chapter enumerates many specific group collaboration practices that they adopted in their first session, which served them well for the remainder of their work together.

- Session *2: The Team Develops Dragging Practices.* Dragging points of geometric figures and observing the consequent changes is a central activity of dynamic geometry. Dragging can be used for a variety of purposes, such as aligning parts of a geometric figure, exploring a construction or testing if dependencies hold during dragging. In their second session, the team developed a number of group practices related to dynamic-geometry dragging.

- Session *3: The Team Develops Construction Practices.* Construction is a conventional focus in learning Euclidean geometry. In this session, the team engages in several traditional construction tasks. In the process, they adopt a series of group construction practices that are specific to dynamic geometry. The chapter also investigates difficulties the team had in constructing figures, how they overcame some of their problems and how they missed opportunities that had been designed into the tasks. During this session, the team displayed significant progress in moving from a visual to a more formal mathematical approach to construction.

- Session *4: The Team Develops Tool-Usage Practices.* In its fourth session, the team honed its skills using the dynamic-geometry tools, including the procedure to define new custom tools. The team adopted additional group practices for using the tools.

- Session *5: The Team Identifies Dependencies.* This chapter explores in even greater detail a particularly exciting developmental breakthrough by the team. Viewed superficially, the team seems to be floundering with a challenging problem involving inscribed triangles. They seem to have digressed even in their collaboration practices. However, in the end of the session it appears that the student who often seems to be the weakest in mathematical understanding solves the problem. The particular geometry task is one that has been used often in the VMT Project and is rarely solved within an hour, even by mathematically experienced adults. A close analysis in this chapter shows how the team actively explored the problem and potential solution techniques through extensive investigation of dragging and construction approaches. The eventual solution actually involved contributions from all three team members and displayed a clear understanding of the solution logic.

- Session *6: The Team Constructs Dependencies.* The team was given another hour-long session to tackle a related dynamic-geometry problem. This time, the triangles were replaced by inscribed squares. The team had not worked with constructing squares before, but eventually arrived at an elegant solution for doing that. Once they constructed the outside square, the whole team immediately expressed knowledge of how to construct an inscribed square in it. This displayed their firm understanding of what they had accomplished in the previous session with the triangles. Their success also confirmed their impressive development of collaboration, dragging, construction, tool-usage and dependency practices.

- Session *7: The Team Uses Transformation Tools.* For their next session, the teacher skipped ahead to an introduction to unrelated tools for rigid transformations (translation, reflection, rotation). Although the team had some success with this topic, they failed to gain much insight into the transformation paradigm of constructing dependencies. Here, analysis revealed the need for considerably more curricular scaffolding, especially supporting enactment of the new tools.

- Session *8: The Team Develops Mathematical Discourse and Action Practices.* The team's final session involved the exploration of different quadrilaterals, to determine dependencies in their construction through dragging. The team investigated seven figures, with very different results. Some figures were too simple and others too difficult to understand through a couple minutes of dragging. However, in working on the second quadrilateral, the team engaged in impressive dragging and in striking mathematical discourse about dependencies. This session displayed both the extent of the team's development along multiple dimensions and the fragility of this development. The analysis of the team's interaction suggests revisions to the curriculum for future research trials.

- *Contributions to a Theory of Mathematical Group Cognition.* In this theoretical chapter, the findings of the preceding analyses are reflected upon as aspects of the theory of group cognition, specifically as applied to school mathematics. The sequences of group practices adopted by the team of students are conceptualized in the light of contemporary cognitive theory. For instance, the group collaborative practices are seen as contributing to a sense of group agency, using insights from Latour and others. The mathematical discourse practices are contrasted to conclusions of Sfard. Group tool-usage practices are considered in terms of Rabardel's concept of instrumental genesis. Dragging is related to embodied group cognition; construction to situated group cognition; and dependencies to designing.

- *Constructing Dynamic Triangles Together.* The concluding chapter has three parts. First, it considers the development of mathematical cognition as a dialectical process rather than a one-time acquisition. Then it recaps the book's implications for re-design of the VMT collaboration environment, especially the curriculum of dynamic-geometry tasks, focusing it even more tightly on dependencies. Finally, it reviews what has been learned from the VMT Project about the development of group practices and suggests prospects for future efforts continuing this research.

Researching Mathematical Cognition

Educators have long felt that developing mathematical cognition was a key to furthering human understanding. For instance, in founding his Academy 2,400 years ago, Plato (340 BCE) insisted that the study of geometry was an important prelude to philosophy. In our own time, computer technology seems to present opportunities for supporting the development of such mathematical cognition by individuals, networked groups and global communities.

The Math Forum (www.mathforum.org) has been providing online resources and services to promote mathematical education since the inception of the Internet (Renninger & Shumar, 2002; 2004). During the past decade, it has conducted the Virtual Math Teams (VMT) research project to explore online collaborative math learning by small groups of teachers and of students. The VMT Project has undergone many cycles of pedagogical design, software prototyping, testing with students and analysis of interaction logs. This research has already been described extensively, including analysis of brief case studies (Stahl, 2006; 2009b; 2013c) (see also www.gerrystahl.net/vmt/pubs.html).

In this book, we take an in-depth look at the interaction of one team of students in order to see *how mathematical cognition develops for that group*. The analysis follows the virtual math team as it engages in mathematical exploration for eight hour-long sessions in a chat room with a multi-user version of dynamic geometry. It describes the display of mathematical reasoning by the team of three eighth-grade female students discussing the dependencies of several dynamic-geometry figures. By analyzing the network of mutual responses displayed in the chat log coordinated with the geometric actions displayed in a session replayer, it is possible to follow the meaning-making processes of the team and to observe how the team learns dynamic-geometry fundamentals, that is, how it adopts a number of relevant mathematical group practices. The analysis is based on displays of evidence of individual cognition, group practices and mathematical reasoning.

The Historical Development of Mathematical Cognition

The contemporary fields of science, technology, engineering and mathematics (STEM), in particular, require a mindset that emerged historically among the ancient Greek geometers (Heath, 1921). Practices of rigor, logical reasoning, causal relationships, lawful behavior, specialized vocabulary and use of symbols are among its characteristics (Netz, 1999). This mindset is a refinement of a more general literacy, representing a qualitative departure from oral culture (Ong, 1998). Many modern citizens have found the transition to this way of thinking insurmountable. A significant number of otherwise well-educated adults readily admit that they are "not good at math" (Lockhart, 2009).

For many people, learning basic geometry still represents a watershed event that determines if an individual will or will not be comfortable with the cultures of STEM. Along with high-school algebra, basic Euclidean geometry—with its notions of dependency and practices of deductive proof—provides a major transition from practical, basic arithmetic to more abstract forms of mathematics. Arithmetic is grounded in counting, which is a common life practice, whereas geometry involves less concrete modes of cognition, which require special enculturation.

Historically, mathematical thinking has been closely associated with science and technology. Thales, the first geometer to formulate a formal proof was also the first scientist in the sense of Western science. Archimedes and da Vinci are other prototypical examples of how math and technology go together.

In the early twentieth century, philosophers sensed a crisis in the foundations of mathematics and science, which had by then become major forces of production in modern society. One of the deepest analysts of this crisis was Husserl, who traced the topic back to "the origin of geometry" (Husserl, 1936/1989). He reflected upon how the early geometers must have built up their field. More recently, Netz (1999) has provided a detailed, historically grounded analysis of the cognitive development of geometry in ancient Greece. He showed how the early geometers developed both a constrained language for speaking mathematically and a visual representation of geometric figures incorporating letters from the world's first alphabet. This was a collaborative achievement, which created an effective medium for communicating about math, for documenting mathematical findings as necessarily true and for thinking about math problems. Geometry was a creation by a small discourse community distributed around the Mediterranean over several generations. They created a system of linguistic, graphical, symbolic and logical *group practices* that became adopted globally and that must be repeatedly adopted by individuals and groups in successive generations.

Mathematics in the sense of geometry and algebra is to be sharply differentiated from practical arithmetic, as long practiced in every civilization. Arithmetic can be mastered by anyone who can memorize the tables and learn a couple of standard manipulation procedures. True mastery of geometry—or more generally of the STEM fields—requires deeper insight or understanding. Attempts that continue to rely exclusively on memorization result in limited success and ultimately in frustration or low scientific self-esteem (Boaler, 2008). The world population is today divided into those people who can engage in mathematical thinking and those who cannot. Mathematical cognition is considered a critical twenty-first century skill of geopolitical import (Looi, So, Toh & Chen, 2011).

Geometry education in schools today has generally evolved in ways that do not feature the cognitive benefits for which geometry was classically valued. Under pressure to teach to tests and to avoid time-consuming exploration, the traditional focus on construction and proof has been largely eliminated (Sinclair, 2008). In particular, the following assumptions—challenged in the research presented here—predominate:

- Learning is treated as an individual mental activity, pursued by students listening, reading, doing homework and taking tests on their own.
- Geometry is taught as a collection of facts: definitions, theorems, procedures.
- Teaching is primarily didactic, with presentations by a teacher and readings in a textbook.
- Geometric figures are presented statically by the teacher or textbook.
- Students seldom construct geometric figures themselves.
- Students are rarely exposed to geometric proofs and deductive argumentation.

In contrast, the approach investigated here features: collaborative learning, student-centered exploration, guided discovery, hands-on manipulation of geometric figures, constructions by students and focus on dependencies as the basis for explanatory proof.

Collaborative learning is not always the best way for everyone to learn everything all the time. However, for many people it is often an effective way to learn certain kinds of things (Stahl, Koschmann & Suthers, 2014). A well-designed collaborative-learning experience can be a powerful way for many people to develop a mathematical mindset and to adopt the mathematical practices that go with it. Considered from the perspective of individual cognition, students might first encounter the insight and understanding that mathematicians need through collaborative experiences within small groups of peers exploring basic geometry. Thereby, they could make the intellectual transition into mathematical literacy practices, which they might otherwise have never attained. Generally, an ideal curriculum orchestrates individual, small-group and teacher-centered classroom activities. The *classroom activity* can provide the larger motivational context and bring in resources, practices and standards from the global mathematical community. The *small-group activities* can provide deep, exploratory experiences with challenging topics requiring new ways of thinking and discussing. Then *individual activities* can help students participate in, individualize, personalize, synthesize, retain, practice and generalize these new ways of thinking.

Collaborative learning is often especially effective when a group confronts a problem that is just beyond the reach of each individual in the group. We shall see that in each of the eight sessions the team of three students accomplishes more in terms of exploring and completing the given task than any one of them could on their own. Not only do they each contribute something to the effort, but their interaction pushes the effort further than the sum of the contributions. In addition to sharing each other's ideas and initiatives, the group process stores up a series of helpful experiences in the form of group practices, which support later efforts.

The same task might stimulate different kinds of learning in different collaborative groups. Novice students could be challenged just to follow specified steps using the named tools. Teams of experienced math teachers could explore alternative approaches to the topic, discuss underlying mathematical relationships or consider difficulties the task might cause their students. A central goal is to engage a group in mathematical discourse that is relevant to their level of mathematical sophistication.

Collaborative learning emphasizes interaction in small groups of learners. The discourse that takes place here brings in notions from community traditions—like the history of geometry—through the use of specialized vocabulary and established tasks. It provides experiences that are motivated by peer relationships and that enable individuals to adopt practices and perspectives from the other group members, from the group's processes and from the larger classroom or cultural community.

Specifically, a carefully designed and guided curriculum in dynamic geometry can introduce groups of students to new math practices, new ways of discussing math and new ways of visualizing mathematical relationships. In this book, we look at data displaying how this can happen. The math practices identified here include group practices of collaboration, dragging, construction, tool usage, dependencies and math discourse. The new ways of discussing math include technical terminology, rigorous formulation, numeric computations, symbolic representations and logical (apodictic or deductive) argumentation. The visualizations include complex interconnected figures, labels, constrained dragging, generalization and variation. These visual props for cognition presage mental visualization capabilities so important to mathematical insight, design and imagination. As we shall see, shared visualizations provide for common ground as a foundation of intersubjective understanding.

One way of judging math understanding is in terms of a sequence of cognitive levels, as proposed by van Hiele (1999). This has been further refined and used by deVilliers (2004) to indicate levels of systematic thinking leading to understanding of axiomatic systems and formal proof. This approach provides a possible way to conceptualize and operationalize the developmental advance of mathematical insight required by STEM.

Another perspective is proposed by Sfard (2008b) in terms of the elaboration of multiple realizations of a mathematical concept. She adopts Vygotsky's insight that a child starts to use a new word before understanding its meaning, perhaps through imitation of another person (or of a text) (Vygotsky, 1934/1986). She then builds on the notion of meaning-as-use (Wittgenstein, 1953). A new mathematical concept gains meaning as it is successfully applied in multiple use applications. Much of what is vaguely referred to as "deep understanding" of mathematics—in contrast, for instance, to rote procedural know-how—can be captured in the idea of multiple realizations: that the mathematical concept can be used, applied or realized in various ways. For instance, a "quadratic relationship" can be represented as an algebraic equation, a graphical curve, a chart of values, a table of differences, a calculus differential, a verbal description. The more ways one has of talking about the concept (thinking to oneself, describing to others, documenting for a community), the deeper ones understanding of its meaning (its appropriate family of applied uses).

In this book, we analyze how a group of three students gradually builds an understanding of geometric relationships, which represents a significant increase in their level of mathematical understanding. They clearly increase the level of their mathematical discourse and analysis in the sense of van Hiele levels, thereby also enriching the multiple realizations of their mathematical conceptualizations. For instance—as we will see in the chapter on Session 3— the students discuss the concept of "dependency" in dynamic geometry during their work on the perpendicular bisector topic in terms of the relationship of perpendicularity in multiple ways: (1) Two lines might have a visual appearance of being perpendicular to each other. (2) The angle between them could be measured to be numerically equal to 90 degrees. (3) The lines could be compared to a prototypical graphical model of perpendicularity. (4) The lines could be constructed to be necessarily perpendicular. (5) One could develop a proof or explanation to show that the lines are perpendicular. Being able to discuss the idea that the dependency of one line's position is dependent on another line's through their mutual perpendicularity in terms of: visual appearance, numeric measurement, comparison with a standard, geometric construction and deductive proof constitutes the beginning of a richly multi-faceted deep understanding. Each of these forms of discussion represents a different van Hiele level. The details of how the team develops these different discourse practices through their online experiences—together with the difficulties they display in these forms of discourse in various interactive contexts—provide insight into the nature of mathematical deep understanding.

This suggests that collaborative online dynamic geometry can provide an effective experience for many students confronted with the challenge of adopting the practices of STEM cognition and discourse. The Virtual Math Teams (VMT) Project has prototyped an approach to this through design-based research. Its goal is to guide teams of students through a multi-dimensional Euclidean "translation" (Stahl, 2013c)—from everyday thinking to mathematical cognition and discourse. It has demonstrated an illustrative concrete curricular approach and provided suggestive evidence of its effectiveness for some student groups. We shall see below how this can play itself out in considerable detail.

The Methodology of Group-Cognitive Development

We would like to see *how* a team of students can develop its group-cognition practices for collaborative dynamic geometry in a VMT environment. If we sense when we look at data from a VMT session that the mathematical cognition of the team of students is developing, how can we determine what the mechanisms of that development are? One of the first questions to address is how best to analyze the generated interaction data to answer this question.

The proper approach to this question involves an "exploratory" study, rather than an attempt to confirm or refute an established theory. We want to document that group-cognitive development is possible by observing a case in which it takes place. We are not interested in a causal generalization, such that a certain condition will always result in (or "predict" with a certain probability) improved learning for students in groups. Such generalizations are probably not the most productive approach in the highly situated contexts of computer-supported collaborative learning (CSCL) efforts in authentic school settings.

As Phillips (2014) recently concluded, educational research should not strive to imitate the controlled-experiment methods of physics or randomized clinical trials in medicine:

> In the hard physical sciences, confounding variables can eventually be controlled, but in research in educational settings, these factors are not nuisances but are of great human and educational significance—control here removes all semblance of ecological validity.... The problem is that in education, just about all the variables are relevant, and controlling them (even if possible, let alone desirable) yields results that are difficult or impossible to generalize to the other almost infinite variety of settings where these variables do, indeed, vary.... Dealing with temperature, pressure, magnetic fields, and the like is one thing; dealing with culture, gender, socioeconomic status, human interests, and the like is quite another. (p. 10-11)

Concerns related to the use of traditional methods from educational psychology—measuring factors affecting individual-learning outcomes—were expressed early in the history of the research field of CSCL, for instance by Dillenbourg, Baker, Blaye and O'Malley (1996, p. 189):

> For many years, theories of collaborative learning tended to focus on how *individuals* function in a group. More recently, the focus has shifted so that *the group itself has become the unit of analysis.* In terms of empirical research, the initial goal was to establish whether and under what circumstances collaborative learning was more effective than learning alone. Researchers controlled several independent variables (size of the group, composition of the group, nature of the task, communication media, and so on). However, these variables interacted with one another in a way that made it almost impossible to establish causal links between the conditions and the effects of collaboration. Hence, empirical studies have more recently started to focus less on establishing parameters for effective collaboration and more on trying to *understand the role that such variables play in mediating interaction.* In this chapter, we argue that this shift to a more *process-oriented account* requires *new tools* for analyzing and modeling interactions. (Italics added)

According to Yin (2009), a case-study approach is the most appropriate method for investigating such a research question (see also Maxwell, 2004; Roth, 2003). We want to see how a group actually goes through a developmental process in order to understand mathematical group cognition. Yin's book is the standard discussion of the use of case studies. In a summary of his text, Yin (2004) writes,

The distinctive topics for applying the case-study method arise from at least two situations. First and most important, the case-study method is pertinent when your research addresses either a descriptive question (what happened?) or an explanatory question (how or why did something happen?). In contrast, a well-designed experiment is needed to begin inferring causal relationships (e.g., whether a new education program had improved student performance), and a survey may be better at telling you how often something has happened.

Second, you may want to illuminate a particular situation, to get a close (i.e., in-depth and first-hand) understanding of it. The case-study method helps you to make direct observations and collect data in natural settings, compared to relying on "derived" data—e.g., test results, school and other statistics maintained by government agencies, and responses to questionnaires.

Accordingly, we adopt a case-study approach for an in-depth analysis of how a group of students increased its mathematical understanding. This is an exploratory study at the small-group unit of analysis, leading to a process-oriented account of group-cognitive development.

In keeping with the exploratory nature of this case study and of its focus on longitudinal development, the analysis will be presented in this book following the chronology of the student interaction. This contrasts with other approaches, which propose theoretical conceptualizations or experimental hypotheses and then assemble evidence from across a dataset to confirm or refute the stated assumptions. The chronological approach has the advantage of being consistent with the sequentiality of the students' own interpretive perspective. The students understand new events in reference to prior actions and experiences—never in relation to events in the future (except, of course, for anticipated responses, goals or events they project). In order to understand how the students make sense of what is going on, researchers should avoid taking into account events from later in the timeline of the data. In order to layer theoretical observations, reflections and conclusions onto a narrative presentation, the analysis will concentrate in each chapter on specific forms of practice that are particularly prominent in the interaction of that chapter's session.

We want to study *development* in Vygotsky's sense. The three students we follow have the same chronological age and the same school level. They have not yet systematically studied geometry. However, they may be at what Vygotsky (1930) calls different "zones of proximal development" in relation to different skill sets. We may, for instance, see that in the group setting, one student can more readily develop her abstract, theoretical reasoning about geometric relationships, while another can more easily develop her facility with dynamic-geometry construction tools and a third can more fluently develop collaboration skills. We want to study how students develop within their zones of proximal development and especially how they extend these zones through their interaction with peers. In particular, for instance, we shall see that their individual skills become shared as the students interactionally appropriate each other's behaviors, insights or skills.

In studies of learning, the unit of analysis can be the *mind* (mental events, psychological states, internal schemas, mental representations, etc.) for individuals, *interaction* (discourse, manipulation of shared figures, positioning of peers, etc.) for small groups or social *practices* (taken-for-given norms, institutions, established genres, etc.) for communities. In group-cognition research, we focus primarily on *interaction*. This is closely related to Conversation Analysis's (Sacks, 1965) focus on informal conversation and Activity Theory's (Engeström, 1999) focus on the workplace activity system. Interaction is mediated by artifacts, tools, language and other resources, so we must look at their roles as well.

We refer to existing theories of geometric cognitive development, including van Hiele's (1999) successive phases of geometric thinking from visual to theoretical, de Villiers' (2004) phases of proof from explanation to axiomatic deduction and Sfard's (2008b) multiple routines of mathematical discourse. We try to see *how* a team of students develops along these general directions through the establishment and adoption of group practices. To carry out our case-study analysis, we focus on the interaction of a specific team.

Early in the VMT Project, we started to do case studies of student interactions. In the VMT SpringFest 2006, there were several teams of students working on problems of mathematical combinatorics. At that time, the VMT researchers directly organized the student teams, grouping together students who volunteered through their teachers. In addition, one researcher sat in the chat room for each session, mostly just to help with any technical difficulties, while trying to avoid interfering with the collaboration or problem solving. Thereby, the VMT researchers had an on-going sense of how each group was doing. Groups B and C in SpringFest 2006 seemed to be doing particularly interesting mathematical work. During the subsequent years, we held weekly data sessions involving the whole VMT research team, in which we went through transcripts and replayings of all five sessions of each of these two groups. This data was the basis for several PhD dissertations, many conference papers and a book (Stahl, 2009b), which included chapters related to those dissertations. While many episodes of interaction by Groups B and C were analyzed, we never discussed the longitudinal development of either group across its sequence of five sessions.

More recently, during the VMT WinterFest 2013, about a hundred students participated in groups organized by several participating teachers (who had taken the semester-long teacher professional-development training offered by the VMT Project during the previous semester). VMT research members monitored the weekly progress of the student teams and communicated with the teachers. In general, no adult was present in these online student sessions. The night after some teams worked on the inscribed-triangles problem (Session 5), we noticed that one of the teams had successfully constructed the figure. We were impressed because that problem usually takes even mathematically inclined adults more than an hour to solve. The team had solved the problem at the very end of their session and had not had much time to discuss the solution or to start to work on the related problems also presented as part of the topic. We emailed the teacher and suggested that she give her class another session to continue work on that topic. We also suggested that she have her students watch a YouTube video on how to use the dynamic-geometry compass tool, which is key to doing the construction.

Soon afterward, the VMT research team looked at that session in our weekly meetings and published an initial exploration of it (Stahl, 2013c, Sec. 7.3). Later, we drafted an analysis of all eight sessions of the team and the research team devoted a two-hour data session to each of the sessions, refining the initial consideration. The analysis indicated that there was rich evidence in the data from this group for a case study of how this group of students developed longitudinally. A workshop was then held at the ICLS 2014 conference (see www.gerrystahl.net/vmt/icls2014), bringing together a number of international educational researchers to deepen the scrutiny of certain issues that were still unclear in the data. The resulting analysis of the selected team provides the basis for the current book.

Our approach used in those analytic investigations is discussed next.

Analyzing Development of Group Cognition

Collaborative learning in small groups has a variety of advantages related to cognitive development. For the participants, it can bring together resources, perspectives and proposals that would not be available to them individually. It can mediate between individual cognition and community knowledge—building group knowledge and group practices that situate community resources and that can subsequently be individuated as personal skills. Thus, it can provide a non-didactic, student-centered, group-constructivist experience, which can overcome some of the customary barriers to effective school-mathematics instruction.

For educational researchers, logs of sessions of student collaboration can provide an intimate view of learning processes as they take place in the media of interaction, which can be captured for detailed study. Through careful design of educational environments, authentic learning experiences can be simultaneously facilitated and systematically documented. The collaborators display for each other their contributions to the group knowledge-building as a necessary and integral part of their interaction, and others (e.g., classmates, teachers, researchers) can observe these displays as well.

This book makes available the displays of a group of students as they learn the fundamentals of dynamic geometry and related skills. In it, one can observe learning taking place as the student team follows an eight-hour trajectory of mathematical topics.

Focus on Group Practices

The aim of the VMT Project has been to iteratively refine an approach to online collaborative mathematics, including elaborating relevant theory, pedagogy and technology. The project has implemented the VMT online environment to support small groups of students working on a sequence of mathematical topics. The VMT software incorporates a multi-user version of GeoGebra, so students can construct and explore dynamic-geometry figures together. GeoGebra (www.GeoGebra.org) is a popular open-source application for exploring dynamic geometry; for descriptions of dynamic geometry, GeoGebra and

VMT, see Stahl (2013c). Guided by an emerging theory of group cognition, the project has evolved a constructivist sequence of dynamic-geometry activities (Stahl, 2012a; 2013b; 2014a; 2014c; 2015). A version of these activities was tried in WinterFest 2013 with over a hundred public-school students.

The VMT project has been driven by continuous cycles of formative assessment directed to the following primary goals:

i.   To facilitate the engagement of student teams in *collaborative knowledge building* and *group cognition* in problem-solving tasks of dynamic geometry.

ii. To increase the quality and quantity of *productive mathematical discourse* by the small groups of students.

iii. To introduce students to the *use of tools* for visualizing, exploring and constructing dynamic-geometry figures.

iv. To develop effective team practices in exploration, construction and explanation of the *design of dependencies* in dynamic geometry.

In order to work together effectively on mathematical topics in the VMT environment, a group of students must increase its ability to collaborate as a team, to engage in effective discourse, to use the tools and features of VMT and of GeoGebra, to decide how to approach stated tasks and to become proficient at analyzing, manipulating and constructing dynamic-geometry figures.

By "*collaborative knowledge building*" or "*group cognition*," we mean the goal of having the students work together and proceed through their session as a team—taking turns, checking for agreement and building on each other's contributions so that the meaning making takes place at the group unit of analysis (Stahl, 2006, Ch. 21). Taking turns chatting or manipulating geometric figures and adopting interactional roles should contribute to maintaining joint attention as a group and shared meaning making, rather than to a division of tasks among individuals.

By "*productive mathematical discourse*," we refer to the quality of the text chat within the VMT environment by a team of students. The team's interaction is considered productive to the extent that it furthers their problem-solving efforts as defined by their current dynamic-geometry task and by accepted mathematical practices—for an illustrative list of math practices, see the US Common Core Standards (CCSSI, 2011). Productive discourse is communication that serves the production of knowledge objects (Damsa, 2014), for instance, text chat in VMT aimed at the group production of a problem solution, a desired geometric construction or a requested explanation—for knowledge-building and knowledge-creation practices, see (Scardamalia, 2002; Scardamalia & Bereiter, 2014). We can view VMT sessions as attempts at both collaborative knowledge building and mathematical knowledge construction. We can identify group collaboration practices that enable knowledge building within the team as well as group mathematical practices that enable the team's construction of knowledge objects like desired dynamic-geometry figures. The group's agency (Emirbayer & Mische, 1998) is oriented to constructing various knowledge objects in the future based on experience in the past and as guided by resources in the present. Textual postings in the chat facility of VMT are often closely associated with the graphical manipulation of geometric objects by the team in the GeoGebra tabs of the VMT interface, and should support such manipulation through guidance, explanation and reflection. The goal of the VMT project is to increase the ability of participating teams to engage in productive mathematical discourse over the lifetime of the teams as they chat in successive sessions (Stahl, 2009b, Ch. 26).

The *"use of tools"* for engaging with dynamic geometry is important because this computer-based form of mathematics requires the ability to select and apply appropriate software tools for each task or sub-task. In the VMT environment, this means knowing how to use the basic tools of GeoGebra. In general, these tools can initially be complicated to use. Some of them require practice. They afford a variety of usages, and students have to "enact" their own styles of using the different tools (Overdijk, van Diggelen, Andriessen & Kirschner, 2014; Rabardel & Beguin, 2005).

The focus on *"the design of dependencies in dynamic geometry"* signifies what we target as the core, underlying skill in mastering dynamic geometry. Figures in dynamic geometry must be constructed in ways that build in appropriate dependencies so that when points of the figures are dragged the dependencies are maintained (invariant). For instance, an equilateral triangle must be constructed in a way that defines and constrains the lengths of the three sides to always be equal; then, even when one vertex of the triangle is dragged to move, rotate or enlarge the triangle, all the sides adjust to remain equal in length to each other. Mastery of dynamic geometry can be defined in terms of the ability to identify effective dependencies in existing figures and to design the construction of such dependencies into new figures to establish and preserve invariants (Stahl, 2013c, Ch. 5).

The disciplinary application of the VMT Project is on introducing teams of students (and teams of their teachers) to dynamic geometry. Dynamic geometry—in our view—differs from previous presentations of geometry in at least three significant features (Stahl, 2013c, p. 63):

    a. *Dynamic dragging* of geometric objects,

    b. *Dynamic construction* of geometric figures and

    c. *Dynamic dependency* in geometric relationships.

The sequence of topics presented to students in WinterFest 2013 was intended to provide experiences in these three features. In this book, we try to observe how the students experience dynamic geometry in their usage of GeoGebra, guided by the instructions in the topics.

In classical Euclidean plane geometry, a point is defined by its original location and it must stay there; however, this location is often taken as arbitrary and an experienced mathematician typically considers how the point could be defined at other locations and what the implications of that might be. Dynamic geometry can be conceived as a computer simulation of what a geometry expert might imagine in this kind of mental variation. Dynamic geometry differs from traditional Euclidean geometry in that a point is not necessarily defined as being at a specific location on the plane. It can be relocated or "dragged" to any other location without changing the relevant mathematics of the situation. This is because the computer software maintains the defined mathematical relationships by moving displayed objects in ways that maintain all defined relationships and confining the movement of the dragged point as necessary to do so. The valid mathematical variations can be visually explored with dynamic-geometry software through dynamic dragging, dynamic construction and dynamic dependencies.

In **Figure 1**, a line segment, a circle and an equilateral triangle have been constructed in GeoGebra using the line segment and circle tools shown in the menu bar across the top. A defining point of each figure has then been dragged and a trace has been created of its positions. Such a trace is typically not visible; rather each student observes the actual motion of the figure on her computer screen in real time. Note that the triangle has been constructed to be equilateral, following the procedure of Euclid's first proposition. While their defining points are dragged, the line segment, the circle and the triangle remain straight, circular and equilateral, respectively.

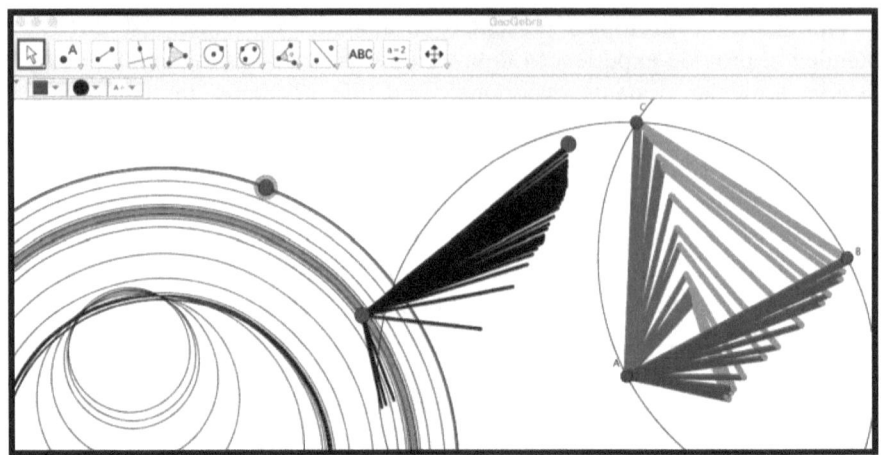

Figure 1. A circle, a line segment and an equilateral triangle are dragged.

By "*dynamic dragging,*" we refer to the multiple roles of the dragging of points and other geometric objects in dynamic geometry (Arzarello, Olivero, Paola & Robutti, 2002). Dragging is not just a way to arrange objects in a static configuration, but rather a way to investigate or confirm relationships in a figure that are invariant under variation of the figure's location (Hölzl, 1996). For instance, when placing a new point at the intersection of two lines, a student should use the "drag test" to confirm that the point cannot be dragged away from that intersection and that if the lines are dragged the point will remain at the re-located intersection. Dragging is also used to investigate conjectures, such as dragging a vertex of a triangle suspected of being equilateral to confirm that the side lengths and angle measures all change together to remain equal. Dynamic dragging represents a different paradigm than the commonsensical static visual appearance of figures—requiring a difficult paradigm shift by students (Laborde, 2004).

By "*dynamic construction,*" we mean that students construct geometric figures in ways that maintain specified relationships dynamically, under dragging. For instance, an isosceles triangle should be constructed with the length of one side defined to be equal to that of another side, not just with the current lengths of the two sides numerically equal. It turns out that the construction procedures presented by Euclid can be used for dynamic construction. This is because Euclid's constructions establish relationships that hold for *any* location of their free points, not just for the particular locations illustrated in a specific diagram. Understanding geometric figures as the results of dynamic constructions provides insight into the necessity of geometric relationships. Dynamic construction represents a different paradigm than the numeric measurement of lengths and angles to determine equality.

"*Dynamic dependencies*" underlie the possibility of dynamic constructions, whose specified characteristics or relationships remain valid under dynamic dragging. A dynamic isosceles triangle ABC (**Figure 2**) maintains the equality of two of its sides, AB and AC, even when an endpoint (B) of one side (AB) is dragged to change its length, because there is a dependency of the length of the second side (AC) on the (dynamic) length of the first side. This dependency may be the result of having constructed point C as a point on a circle centered on point A and defined as passing through point B. As long as point C remains on this circle, no matter how any point is dragged (changing the locations and sizes of the circle and line segments), the lengths of sides AB and AC will remain equal because they are both radii of the same circle. The invariance of the isosceles triangle (that its two sides are always of equal length) is designed, constructed, enforced and explained by the dependency of the side lengths on the circle. The ability of a team to design dynamic dependencies requires the development of a variety of group-cognitive, mathematical and group-agentic practices, including thinking, speaking, analyzing, explaining and constructing in terms of dynamic dependencies.

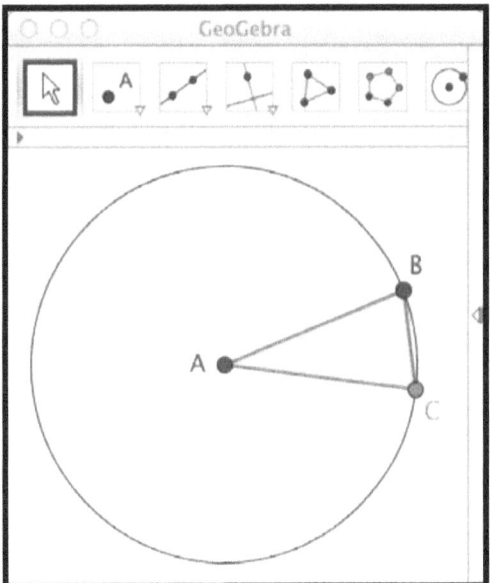

Figure 2. Dynamic isosceles triangle

The paradigm of dynamic geometry is quite subtle—both in itself and in its comparison to Euclidean geometry. Understanding how things work and move involves a number of insights. Being able to design dynamic constructions and predict how the figures will move or what will remain invariant is complicated.

We will see that the students we study adopt a number of group practices in their effort to accomplish dynamic-geometry tasks and we will see that their resulting mastery is both impressive and fragile in specific ways.

This book will follow one virtual math team through its eight online sessions in early 2013, in order to see how this team started and progressed. The analysis will focus on the development by the team of group practices supporting:

i.   Its effective team collaboration,

ii. Its productive mathematical discourse,

iii. Its enacted use of dynamic-geometry tools and

iv. Its ability to identify and construct dynamic-geometric dependencies by:

    a.   Dynamic dragging of geometric objects,

    b.   Dynamic construction of geometric figures and

    c.   Dynamic dependency in geometric relationships.

We focus on "*group practices*" as foundational to collaborative learning. This is in keeping with the "practice turn" in contemporary social theory and epistemology (Schatzki, Knorr-Cetina & Savigny, 2001). According to Reckwitz (2002), a practice is "a routinized type of behavior which consists of several elements, interconnected to one another: forms of bodily activities, forms of mental activities, 'things' and their use, a background knowledge in the form of understanding, know-how, states of emotion and motivational knowledge" (p. 249). Social practices form our background, tacit knowledge. This is the alternative to rationalist and cognitive philosophies, as proposed by Heidegger (1927), Wittgenstein (1953) and Polanyi (1966).

Turner (1994) situates practice theory as a critical reaction to prominent contemporary social theories: cultural mentalism (economism and cognitive psychology), textualism (20th century French philosophy) and intersubjectivism (esp. Habermas). Practice theory is most extensively propounded by Bourdieu (1972/1995). He uses the term "habitus" for our systems of durable, transposable dispositions—or organization of conventionalized, routinized, objectified and embodied habits. They are life's lessons, whose origins in specific lived experiences are long forgotten. They are second-nature ways of behaving, speaking, moving and thinking. Due to their generalized and tacit nature, they allow people to say more than they consciously know.

As with other concepts, in the theory of group cognition we construe practices primarily at the small-group unit of analysis, rather than as habits of individual bodies or cultural conventions of whole communities—in contrast to Bourdieu and his followers. The group practices are what makes collaboration possible: "The homogeneity of habitus is what—within the limits of the group of agents possessing the schemes (of production and interpretation) implied in their production—causes practices and works to be immediately intelligible and foreseeable, and hence taken for granted" (Bourdieu, 1972/1995, p. 80).

Because the meaning of group practices is understood the same by all group members, the members can understand each other's actions and their references to those actions. The intersubjectivity of the group is based on this *shared meaning*. The sharing of meaning is a product of the group interaction: it is produced in the interaction (as opposed to originating in the minds of the individual group members). Practices are proposed—whether verbally or in action—and then discussed, negotiated, accepted, put into regularized practice, generalized across instances of practice and incorporated into the group's habitus. Henceforth, it is accepted within the group interaction without need for explicit mention or questioning. In fact, it is more visible in its absence. Through the process of group meaning making, the meaning of the practice is established as the same—for all practical purposes—in the understanding of all group members. This does not imply that each member would be able to articulate the same expression of meaning if interviewed or that this meaning is somehow similarly represented in each member's mind. It does entail that group members will naturally respond appropriately to occurrences of the practice within the group interaction. This shared tacit understanding—and not some recursive form of explicit verbal agreement—is the basis for common ground and intersubjectivity in the group (Stahl, 2016a).

Our group-level analysis of practices avoids the critique of practice theory in Reckwitz (2002). Reckwitz ends up assuming a mentalist view and then cannot reconcile the sense of shared understanding associated with social practices with mental agreement among individuals. But for us, the students in a team are co-present in a shared world when they engage in shared practices and understand them tacitly (Stahl, Zhou, Çakir & Sarmiento-Klapper, 2011). Reckwitz also does not have an analysis of how practices are acquired and is consequently worried about relativism. However, we can see in the student displays how they negotiate, refine and adapt group practices.

Within the VMT context, Reckwitz's issue of relativism—e.g., that a team might reach results that are not correct according to conventional mathematical standards—is addressed at several levels. First, the curriculum guides students in the direction of conventionally established mathematical results and poses questions to check on their progress. Second, the GeoGebra software implements the definitions and axioms of dynamic geometry, so that attempts to construct and drag figures provide extensive feedback concerning what is mathematically valid. Third, after a session is finished, a teacher checks the student conclusions and may discuss them in class. Fourth, a central goal of the project is to guide students to higher levels of geometric sophistication in the direction of axiomatization, which is the standard for truth in mathematics. Perhaps most importantly, the collaboration process itself provides a check in that a student proposing a step in a solution must convince the others in the group; while none of the students knows the "correct" solution in most cases, they are all involved in assessing the arguments for each deductive step. In the VMT curriculum, there is generally not a single correct answer, but alternative problem-solving approaches; tasks and instructions are relatively open-ended and student creativity is encouraged. Nevertheless, mathematical standards of validity are structurally included in the project activities, the math technology and the overarching pedagogy.

The development of acceptable mathematical practices is a central goal of the VMT project. The enactment of group mathematical practices by a team is not only guided by curriculum and checked by community standards; it is mediated by the available technological tools (Carreira et al., 2016). In the analyzed sessions, students approach a given curricular challenge with the tools and practices of GeoGebra, as enacted in their previous work and learning. They construe the problem and solution strategies in terms of possible GeoGebra constructions and practices. This approach of considering potentially useful GeoGebra tools and previously effective construction procedures is itself a group practice that the team has to adopt at some point. As we will see in session 3, in working on problems of perpendicularity, it took awhile before the student team started to apply the tools and practices they adopted in their previous topic to the current topic's challenge. First, they tried to use practices of visual shape and measurement before they turned to group construction practices from their construction of the equilateral triangle.

The "affordances" (Gibson, 1979) of the GeoGebra technology are defined largely by its software tools (such as the construction tool to produce a line segment between two points or the attributes tool to change the color of an object). However, these affordances must be enacted by the users as they adopt practices that apply these tools in practice—such as first defining the two points with the point tool, then finding and selecting the segment tool and then clicking on the points one after the other to construct a segment between them. It is the collection of practices adopted by a user for using the tools that define the affordances of the technology for that user. The group "enacts" (Weick, 1988) the tool by adopting the corresponding group practices. This involves the exercise of collaboration and discourse practices, as well as those related to geometric construction. Therefore, it is important for us in analyzing the development of the team's mathematical group cognition to identify the various kinds of practices involved, such as group collaboration practices, group dragging practices, group construction practices, group tool-usage practices, group dependency-related practices and group mathematical discourse or action practices.

The goal of identifying group practices as they are adopted by the student team raises the question of how to conduct an appropriate fine-grained analysis of interaction. We turn next to that.

Sequential-Interaction Analysis

Methods of evaluating how small groups learn when interacting through computer-supported collaborative learning (CSCL) systems are not well established (Stahl, Koschmann & Suthers, 2006). In particular, the most common methods—inherited from research in educational psychology oriented to individual learning outcomes—involve coding and aggregating utterances, which generally eliminate the important sequential structure of the discourse. The resultant statistical computations can provide comparative measures of outcomes, but analysis of concrete sequences of discourse is often needed to reveal the mechanisms or the group processes involved in producing such outcomes. Sequential analysis of informative interactions can often show in more insightful detail how specific support functionality in CSCL software is effective in mediating productive group work. For this, we need to focus analysis on the sequential structure of the interaction as a meaning-making process.

A primary concern for designers of educational interventions should be the extent to which groups using their approach are actually supported in the ways intended by the design of the intervention. Determination of what learning does and does not take place in the environment and the role of specific technical or curricular functionality in supporting or failing to support that learning is essential to re-design for subsequent iterations of the development cycle.

The VMT Project is a *design-based research* effort, which means that it undergoes cycles of design, implementation, testing, evaluation and re-design (DBR Collective, 2003; Stahl, 2013c, Ch. 11). The project has gone through countless design cycles during the past decade, systematically evolving a CSCL environment for small groups of students to learn mathematics together. In particular, the designers of the VMT environment have developed software, curricular resources, teacher-professional-development courses and best practices to introduce students to the core skills of dynamic geometry. Project staff members need periodic feedback on how their prototypes are succeeding in order to redesign for improved outcomes.

The question addressed by this book is: How well did students in the WinterFest 2013 iteration of the VMT Project learn the skills that the environment was intended to support? The point is not to come up with a rating of the success of this approach, as though the software, curriculum and pedagogy were in a final state. It is also not to compare how users "feel" or whether they "succeed" when using VMT versus not using this support system. Rather, the aim is to observe just *how* teams of students learn targeted skills or how they fail to learn them within the designed environment. In other words, what group practices did the team adopt that helped them master the tasks and what practices seem to still be missing from their work that might have helped them? These observations should be concrete enough to drive future cycles of re-design.

The driving question of the VMT Project can be formulated as: "How should one translate the classic-education approach of Euclid's geometry into the contemporary vernacular of social networking, computer visualization and discourse-centered pedagogy?" (Stahl, 2013c, p. 1). The approach is to use a computer-based form of geometry known as dynamic geometry (e.g., Geometer's Sketchpad, Cabri, GeoGebra). More specifically, the project is based on the principle that "the key to understanding dynamic geometry is not the memorization of terminology, procedures, propositions or proofs; it is *dependencies*" (p. 11). That is, the intention of the VMT Project is to support teams of students to develop their ability to identify dependencies in geometric figures and to use those dependencies in their own construction of similar dynamic-geometry figures.

In the beginning of 2013, the Math Forum sponsored a "WinterFest" in which teams of three to five students participated in a sequence of eight online sessions using the VMT environment. The groups were organized by teachers who had completed a semester-long teacher-professional-development course in collaborative-dynamic-mathematics education, offered by Drexel University and Rutgers-Newark. The VMT environment at that time included the first multi-user dynamic-geometry system, an adaptation to VMT of the open-source GeoGebra software. The mathematical topics for the eight sessions were embedded in multiple tabs of VMT chat rooms for each of the sessions. The topics were developmentally designed to gradually convey an understanding of geometric dependencies.

In order to observe in sufficient detail how a group of students learned over time to work on dynamic geometry and to identify geometric dependencies, the VMT Project staff held weekly "data sessions" (Jordan & Henderson, 1995) in which they looked at logs of the sessions of a particular group of three students, who called themselves Cheerios, Cornflakes and Fruitloops. The members of this "Cereal Team" were 8th grade girls (about 14 years old) in an after-school activity at a New Jersey public school. They were beginning algebra students who had not yet taken a geometry course. This particular team came to the VMT project team's attention in connection with their performance on Topic 5. This challenge had been worked on by many groups during the VMT Project and had become a useful benchmark for observing groups identifying geometric dependencies (Stahl, 2013c, Ch. 7). The selection of a group that did interesting work on Topic 5 allowed the VMT researchers to generalize by understanding the Cereal Team in the context of parallel work by other teams of various composition.

Figure 3 shows the VMT application near the end of the Cereal Team's Session 5. On the left is the GeoGebra panel, as it appeared for each student. On the right is the chat panel, where the three students synchronously communicated textually. The small squares embedded in the chat indicate GeoGebra actions. The VMT interface and functionality are described in (Stahl, 2009b, Part IV) and the multi-user version of GeoGebra is discussed in (Stahl, 2013c, Ch. 6). The larger set of inscribed triangles (ABC/DEF) was displayed originally as part of the task. The task, described in the text box, includes exploring that figure. The smaller triangle, GHI, with the circles defining points M and R, were constructed by the students, as described by Cheerios in the chat.

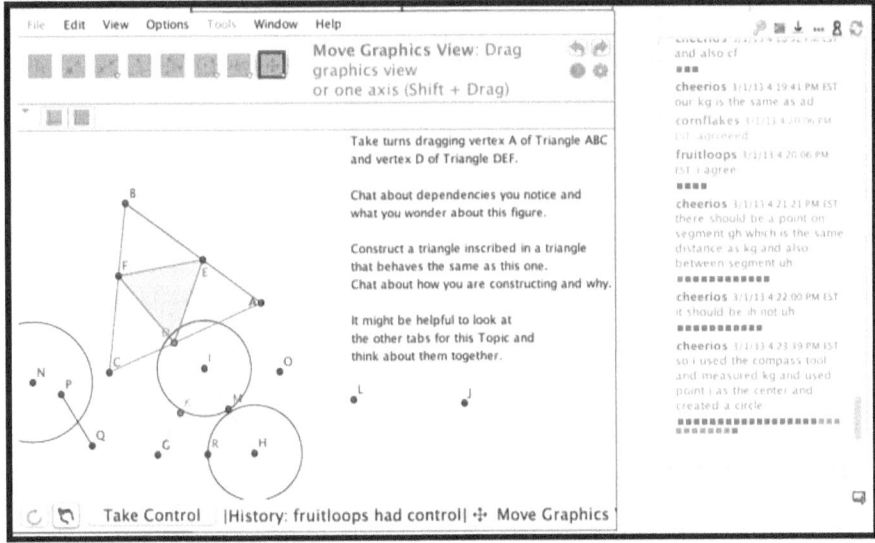

Figure 3. The Cereal Team works on Topic 5.

During WinterFest 2013, 34 teams of middle school and high school students each worked on eight dynamic-geometry topics in the VMT online environment, supervised by their 10 teachers. Based on a review of logs conducted by the VMT researchers with input from the teachers involved, of the WinterFest 2013 teams, the Cereal Team consisting of Fruitloops, Cornflakes and Cheerios was selected as "the most collaborative team," earning it first prize in WinterFest 2013. By the end of the WinterFest, 148 students had participated in at least 7 sessions; they were all awarded prizes for the sustained involvement of their teams.

An initial analysis by the VMT research team of the Cereal Team's first session (on Topic 1) demonstrated how much they had to learn about collaborating, chatting, navigating VMT, using GeoGebra tools, arguing mathematically, manipulating dynamic-geometry objects and constructing figures (Stahl, 2013a). A more detailed study of their logs for sessions 5 and 6 revealed considerable progress, but still showed some major holes in their understanding of how to design the construction of dynamic-geometry figures to incorporate specific dependencies (Stahl, 2013c, Ch. 7). Another paper analyzed their work on their final session (Stahl, 2014b). Synthesizing these studies of individual sessions, elaborating them more fully and filling in the missing sessions to provide a longitudinal analysis, the present book reviews the team's progression through all of its eight sessions.

Although design-based research is a popular approach to the development of educational software, especially in CSCL and Technology-Enhanced Learning, there is little agreement on how to evaluate trials in a way that contributes systematically to re-design. The theory of Group Cognition proposed that one could make collaborative learning—or group cognition—visible (Stahl, 2006, Ch. 18), based on the principles of ethnomethodological description (Garfinkel, 1967). This is because meaning making is an intersubjective or small-group process, requiring group members to make their contributions visible to each other, and therefore also to researchers (Stahl, 2006, Ch. 16). As the editor's introduction to (Garfinkel, 2002) explains, "the sounds and movements that comprise social action are meaningful creations that get their meaning from the shared social contexts of expectation within which they are enacted…. Intended meanings, however, can only be shared if they can be successfully displayed before others in expected ways" (p.57).

This book's analysis of the meaning-making process focuses on the sequential response structure (or "adjacency pairs") of utterances, which build on previous utterances and elicit further possible, anticipated or expected responses (Schegloff, 2007). The analysis re-constructs the web of situated semantic references: "The meaning of the interaction is co-constructed through the building of a web of contributions and consists in the implicit network of references" (Stahl, 2009b, p. 523).

The adjacency pair is the smallest unit of analysis of collaborative meaning making. An isolated utterance of an individual does not have well-defined meaning. Its meaning is defined by its use in communication—as a response to a prior utterance of someone else and/or as an elicitation of a future response of someone else. The elicitation-response structure of an adjacency pair defines a minimal inter-action between two or more actors. The meaning of the eliciting utterance is defined in large part by the response (as situated in the larger context), which takes it up in a specific way and thereby grants it its meaning. Often, the meaning granted by the response is not disputed or further negotiated, although it may become clarified, altered or amended as the interaction continues. Sometimes, the establishment of shared meaning through the adjacency pair is problematic. This is typically signaled by the utterance of a repair move by one of the actors, resulting in a continuation of the adjacency pair and its meaning-making process.

Note that meaning is constructed by more than one individual through the elicitation-response pair. That is why interaction analysis is considered to take place at the small-group unit of analysis. If one attributed the meaning of a single utterance to the mental state of the individual making the utterance, than that would be an analysis at the individual unit—and would imply some form of access to the individual's mental state. Single utterances can rarely be adequately interpreted in isolation; they typically include indexical elements that reference prior utterances and other elements of the interactional situation (Zemel & Koschmann, 2013). Therefore, they must be analyzed in terms of their sequential position with respect to utterances of other people.

Most published sequential analyses of conversation are limited to brief excerpts; this book's analysis of each hour-long session—especially considered within the larger context of the VMT Project—goes beyond the analysis of even so-called "longer sequences" (Stahl, 2011d) toward longitudinal analysis of collaborative learning across multiple sessions. We want to observe the collaborative learning of the team as it evolves during eight hours of intense, complex interaction.

Analysis of longer sequences is more important in studying geometry instruction than in most conversation analysis. While ethnomethodologically informed Conversation Analysis (Garfinkel, 1967; Goodwin & Heritage, 1990; Sacks, 1965; Schegloff, 2007) is interested in how meaning is socially constructed in the momentary interaction, we are here concerned with both (a) longer chains of meaning making and (b) how the meaning-making process itself changes as the group learns to collaborate and to engage in mathematical discourse.

(a) Perhaps geometry's greatest contribution to the development of human cognition was to systematize the building of *chains of reasoning*—presented as deductive proofs or specially structured constructions of graphical figures (Latour, 2008; Netz, 1999). Euclid's proofs could extend to over forty steps, each specified in a prescribed technical language and accompanied by a diagram representing a correspondingly complicated construction. The cognitive capacity to follow—let alone to invent—such a sequence of deduction or construction required the development of meta-cognitive planning and agentic regulation skills (Charles & Shumar, 2009; Emirbayer & Mische, 1998; Stahl, 2005). These skills have since the time of the early Greek geometers become ubiquitous in literate modern society (Ong, 1998). They underlie our scientific worldview and technological lifestyle. Sophisticated planning skills have become second nature (Adorno & Horkheimer, 1945) to us and we now assume that people are born with rational skills of planning and arguing. It has taken seminal studies of philosophy (Heidegger, 1927) and psychology (Suchman, 2007) to dispel the common rationalist assumption (Dreyfus, 1992) that our actions are the result of previous mental planning, rather than that reasoning is generally posterior rationalization (Stahl, 2013c, Ch. 3), and that we must learn how to make up these explanations after our actions as little retroactive stories (Bruner, 1990), in order to understand and justify them. We would like to see how a young, novice team could develop such sequential reasoning skills, guided by experiences involving geometric construction, analysis and planning. We hypothesized that studying geometry can be an occasion during which significant steps of learning about deductive reasoning can take place. We shall look for the Cereal Team's adoption of group practices that involve group agency of sequences of task steps.

(b) Following a development of group agency over time involves the longitudinal analysis of longer sequences of interaction or comparison of excerpts at different points in a temporally extended learning. Analysis of a single moment can reveal how participants take their activity as instructional or can display signs of having learned something new (Koschmann & Zemel, 2006; Zemel, Çakir, Stahl & Zhou, 2009). However, it can be more informative to compare and contrast interactions at different times to reveal how groups and their participants have taken up previous experiences in current interaction (Sarmiento & Stahl, 2008a) and *how that makes a difference to their current meaning making*. We would like to observe the evolution of group practices and individual skills or understandings over time. Our analytic goal can be called a "learning trajectory." Such a trajectory has been characterized as follows:

> A researcher-conjectured, empirically supported description of the ordered network of constructs a student encounters through instruction (i.e., activities, tasks, tools, forms of interaction and methods of evaluation), in order to move from informal ideas,

through successive refinements of representation, articulation and reflection, towards increasingly complex concepts over time. (Confrey et al., 2009, p. 346)

Note the central role of instruction here. Instruction is conceived here as the provision of a carefully designed learning environment. As Lehrer and Schauble (2012) put it, "The benchmarks of learning tend not to emerge unless someone carefully engineers and sustains the conditions that support them" (p. 705). We will see how the VMT environment guides the student team's learning trajectory as they adopt group practices that enable them to refine their representations, articulations and reflections over time.

Temporal analysis of interaction in the VMT setting raises some special concerns related to the identification of adjacency pairs in discourse (Zemel & Çakir, 2009). In everyday conversation, an utterance typically responds to the immediately preceding utterance, and the previous speaker generally listens silently as the new utterance is produced. However, in text chat people can be typing simultaneously. In the VMT chat system specifically, there is an awareness message indicating who is typing but one cannot see what is being typed until it is posted as a finished message. People do not always refrain from typing a new message while they see that someone else is typing. Sometimes, someone may delete a message they started to type without posting it because of what someone else meanwhile posted. In any case, a new posting usually does not respond to a previous posting unless that posting was completed before the new one was starting to be typed. Therefore, it may be best to refer to the interaction as being structured by "response pairs" rather than "adjacency pairs," where a response to a previous posting may not be immediately adjacent to it. In addition, in the VMT environment, a posting may be responding to the dragging of a geometric object or a construction action in a GeoGebra tab. The students generally display information to enable their partners to follow the intended response structure by the way they design their posting—and researchers can also take advantage of these displays. Although information about GeoGebra actions is not generally included in the chat log excerpts as presented in this book, all of it is available to researchers (as well as to the students and their teachers) in more detailed spreadsheets and files for replaying sessions. The analyses in this book take into account the full interaction data in order to analyze carefully the online response structure and group meaning making displayed there.

The Display of Collaborative Development

Learning is traditionally conceived as a change in testable propositional knowledge possessed by an individual student (Thorndike, 1914). Opening up an alternative to this view, Vygotsky argued that students could accomplish epistemic (knowledge-based) tasks in small groups before they could accomplish the same tasks individually—and that much individual learning actually resulted from preceding group interactions (Vygotsky, 1930), rather than the group being reducible to its members as already formed individual minds. This called for a new conception of learning and social interaction. Vygotsky conceived of group interactions as being mediated by artifacts, such as representational images and communication media. His notion of artifact included physical tools, conceptual signs and spoken language. More recently, educational theorists have argued that student processes of becoming mathematicians or scientists, for instance, are largely a matter of mastering the linguistic practices of the field (Lave & Wenger, 1991; Lemke, 1993; Sfard, 2008b).

Paradigms of learning focused on individual minds require methodologies that test individual changes over time and interpret them in terms of some theory of mental processes that are not directly observable—such as mental models, mental representations, cognitive change, cognitive convergence or cognitive conflict (Stahl, 2016b). In contrast, a view of learning focused on group interaction can hope to observe processes of group cognition more directly. A reason for this is that in order for several students to collaborate effectively, they must display to each other what the group is planning, recalling, doing, concluding and accomplishing. These displays can take place in the physical world through speech, gesture and action or in the virtual world through text chat and graphics, for instance. They are in principle visible to researchers as well as to the original participants.

In practical terms, it is often difficult for educational researchers to capture enough of what is taking place in group interactions to be able to reliably understand what is going on as well as the participants do. Capturing face-to-face collaborative interaction in an authentic classroom involves many problematic complications, including selecting video angles, providing adequate lighting, capturing multiple high-quality audio recordings, audio-to-text transcription, representation of intonation or gesture and synchronization of all the data streams (Suchman & Jordan, 1990). In this book, we present data that was automatically captured during an online chat involving three students. All of their communication and action that was shared within the group is available to us as analysts in exactly the same media and format as it appeared for the students, as well as in automatically generated, time-stamped textual logs. So the issues of selection, coordination, interpretation, partiality and representation of audio, articulation, gesture and other data are not problems here the way they are in recordings of face-to-face settings. In particular, all the representational graphical images and textual language shared by the group are available in detail to the researchers in their original formats. We can use methods of interaction analysis, video analysis or conversation analysis (Jordan & Henderson, 1995; Koschmann, Stahl & Zemel, 2007; Schegloff, 1990), adapted to our online math-education setting (Zemel & Çakir, 2009).

Of course, interpretation and analysis of meaning can still be controversial in our approach. However, the raw data is available and excerpts of it can be included easily in the presentation so that readers can see where interpretive decisions have been made and can judge for themselves the plausibility of the analysis. There are no hidden stages of imposing categories and theoretical perspectives on the presented data. The students speak in their own words (including punctuation and typos). Furthermore, the data displays the words and actions of the students as they are engaged together, rather than providing retrospective statements of the students in reaction to questions when they are no longer engaged in their mathematical tasks or situated together online (as in surveys, questionnaires, focus groups, interviews, reflection papers or pre/post-tests).

For some time, we have proposed the idea of focusing on the small group as the unit of analysis and foregoing any reliance on theories of mental processes in favor of observing the visible interactions (Stahl, 2006, Ch. 21). We spent a decade developing an online environment which could support collaborative learning of mathematics and also be instrumented to capture group interaction (Stahl, 2009b, Part IV). Our research and theory now distinguish distinct learning processes at the individual, small-group and community units of analysis (Stahl, 2013c, Ch. 8). Although we recognize that these processes are inextricably intertwined in reality, we focus methodologically in this book on the small-group unit of analysis, which is where individual learning, group becoming and community practices are generally most visibly displayed (Stahl, 2016b). One ends with a similar analytic approach if one adopts the related theoretical perspective of commognition (cognition as communication) (Sfard, 2008b) or dialogicality (Wegerif, 2007): individual thinking, group interaction and community knowledge building are all a matter of discourse, which is fundamentally intersubjective and analyzable at the small-group unit of discourse or interaction, rather than in terms of mental representations.

In particular, we identify *group practices* that the team adopts in their interaction. For instance, in the analysis of the team's Session 1, we identify a number of group collaboration practices, and in the analysis of other sessions, we identify various kinds of group mathematical practices. These practices often appear in the discourse of the group and can be considered linguistic practices; cultures are often identified by their language and small groups sometimes develop or adopt distinctive methods of communicating.

Group practices are like social practices. Whereas social practices are common ways of interacting shared by members of a culture, group practices are routines adopted by a small group and taken as understood the same by the members of the group. The concept of practices (or member methods) is borrowed from Ethnomethodology (Garfinkel, 1967; Stahl, 2012b). Ethnomethodology is a phenomenological approach to sociology that tries to describe the methods that members of a culture use to accomplish what they do, such as how they carry on conversations (Sacks, Schegloff & Jefferson, 1974) or how they "do" mathematics (Livingston, 1986). In particular, the branch of Ethnomethodology known as Conversation Analysis (Sacks, 1992) has developed an extensive and detailed scientific literature about the methods that people deploy in everyday informal conversation and how to analyze what is going on in examples of verbal interaction. Methods are seen as the ways that people produce social order and make sense of their shared world (Stahl, 2011c). If members of a linguistic community did not have shared understandings and practices, they could not understand each other's words or behaviors. As we shall see, group practices in online settings are often structured differently than corresponding face-to-face practices, even if they must accomplish similar functions.

The data for this book consists of eight hours of online interaction by a team of three students. This data provides a rich source for analysis of collaborative learning of dynamic geometry. The learning of geometry has been a pivotal moment in the cognitive development of many people and of humanity generally, but also a difficult achievement for many people (Lockhart, 2009; Sinclair, 2008; Stahl, 2013c, Ch. 1). As noted earlier, learning geometry involves the kinds of abstract, rigorous, systematic, argumentative discourse practices that are common to scientific, technical, engineering and mathematical (STEM) work. In the interactions of this team of students, we can observe how they develop many skills and practices important to collaborative learning and to doing mathematics.

Perhaps most importantly from a research perspective, the student team displays its learning in its chat discourse and in its geometric actions. In the displayed group learning, we can see how progress in collaboration, math discourse and dynamic geometry comes about. The goal of the following analysis is therefore to let the students' voices speak for themselves and to observe what the students display to each other—especially as they are establishing new group practices.

One qualification to this ideal is that making sense of math discourse often requires the analyst to be aware of the mathematical affordances of the curricular topic being discussed (as for Livingston, 1986). The student appropriation of the topic results from the dialectic between the group agency of the team situated in its activity system and the topic invested with the intentions of its designers (Overdijk et al., 2014). For this reason, we will sometimes describe the intended lesson of a curricular resource in order to understand how the team took advantage of a potential learning opportunity by bringing the resource into use in a specific concrete way. This will also help us to note student actions that deviate from usual approaches or even go beyond the expectations underlying the topic designs.

The WinterFest series of online sessions is designed to be an educational experience, requiring students to engage with certain mathematical content. Since there is typically no teacher present in the chat rooms during these collaborative sessions, the facilitation role is largely "scripted" (Fischer, Mandl, Haake & Kollar, 2006; Kobbe et al., 2007) by the texts in the chat room tabs. The analysis will show how the team interpreted or reacted to this guidance and enacted or applied the topic's instructions. This will provide feedback on how successful the design of the topic was.

The goal in this book is to observe the student displays of how a particular virtual math team learns to collaborate, discuss mathematics and engage in dynamic geometry during WinterFest 2013. We try to observe the learning as it takes place. We systematically document this learning in terms of the team adopting group practices. The VMT instrumentation allows each session to be replayed, so one can see the same thing the group of students saw. By studying the interactions in which students display their emergent understanding to each other, one can see the collaborative learning taking place. This makes available to researchers not just occasional pre and post states, but the on-going problem-solving and knowledge-building processes as they unfold at the group unit of analysis.

We now proceed sequentially through the eight sessions of the Cereal Team. Two sessions were held each week from February 15 to March 11, 2013. We will be trying to see how the team improves:

i.   Its effective team collaboration,

ii.   Its productive mathematical discourse,

iii.   Its enacted use of dynamic-geometry tools and

iv.   Its ability to identify and construct dynamic-geometric dependencies by:

     a.   Dynamic dragging of geometric objects,

     b.   Dynamic construction of geometric figures and

    c.   Dynamic dependency in geometric relationships.

We will observe how each of these progresses as the team interacts during its sessions. Much of the text chat is reproduced verbatim, as typed and posted by the students, in the ensuing chapters. The analysis is based primarily on observing the sessions as they unfolded, through the VMT Replayer application. Occasional screen images from the Replayer are included in the presentation of the analysis. The Replayer and the data files for the sessions are freely available for download at: http://gerrystahl.net/vmt/icls2014. Complete logs are also available in spreadsheet files there, which list all the GeoGebra actions as well as all the chat postings.

Session 1: The Team Develops Collaboration Practices

In this and the following chapters, we will chronologically consider the eight hour-long sessions of the Cereal Team, reviewing most of their chat postings and much of their GeoGebra activity. This will present their whole VMT experience in considerable detail.

In each chapter, we will focus on one dimension of their collaborative learning. For instance, in this chapter, we will concentrate on *how the team learned to collaborate effectively*. They started their first session with little sense of how to interact in this environment, went through a series of episodes and gradually developed a set of *group collaboration practices*. We will try to document how their stages of learning to collaborate are displayed in their interaction. In subsequent chapters, we will notice how the team's collaboration practices are further developed, but will concentrate then more on the team's mathematical development.

Learning is a gradual process, not a binary switch: first ignorance and then suddenly knowledge. A group of people can learn to work and think together, but this involves a prolonged process of repeatedly accomplishing a whole complex of advances. To become an effective collaborative team, a group of people must establish mutual co-presence in a communal world and they must develop intersubjectively shared understanding before they can engage in effective group cognition (Stahl, 2016a). Tomasello (2014) provided a theory of how the human species evolved its distinctively human ability to collaborate in the physical world and to establish cultural practices during the past 400,000 years. The Cereal Team had to do something analogous within an unaccustomed online setting in less than an hour. Of course, they already knew how to interact face-to-face, so they had a lot of experience to build on and they could carry over certain components from one environment to the other, such as their mutual trust, established personal relationships and common ground as school acquaintances.

In virtual (online) contexts, groups of people have to create practices of interaction and communication which are different from those in face-to-face settings, but which are analogous in meeting the same basic needs for accomplishing tasks together (Zemel & Çakir, 2009). In this chapter, we shall observe in the logs interactional displays in which the Cereal Team proceeds through processes of learning to collaborate. We will identify and highlight a sequence of events in which the team creates and adopts group practices that will subsequently enable it to engage in collaborative learning of dynamic geometry within the VMT environment.

The sequence of practices actualized by the team in this session does not constitute a general theory comparable to Tomasello's. This is a single case of highly situated interactions, which cannot be taken as either systematic or generalizable. However, the researchers who analyzed this data have been observing similar groups for a decade or longer, and can recognize practices that are typical under such circumstances. The following analysis shows how a group of collaborating students can make visible through their interactive displays characteristic ways that students learn to collaborate in CSCL settings. Individual interaction episodes become codified as group practices through the team's agreement on them and repeated usage.

Normally, the adoption of group collaboration practices involves the interplay of phenomena at the individual, small-group and community levels— such as the appropriation of a community resource (e.g., a prompt in the curriculum or a technical term/concept from geometry) by an individual participant, which is then negotiated and adopted by the group.

In his evolutionary analysis, Tomasello (2014, p. 78) emphasizes the dual-level structure of collaboration: simultaneous jointness and individuality, where each participant jointly attends to joint goals, but from her own individual perspective. In the following data, there is a strong sense of individual perspectives, which not only interact intimately to achieve team goals, but are also gradually mutually adopted by the different students. Furthermore, the cultural conceptions—such as the mathematics community's definition of 'dependency'—affect the team and its members' individual understandings. Tomasello argues that the prehistoric evolution of joint intentionality within cooperating small groups of proto-humans led over hundreds of millennia to our distinctively human forms of collective (community) intentionality with language, culture and eventually social institutions. This provided a basis for and eventually resulted in the development of modern individual intentionality, with its senses of subjectivity, reflection, rationality and objective reality. In this analysis, group cognition was foundational for both individual cognition (thinking) and community cognition (culture). By focusing on team interaction in the VMT data, we shall observe some of the collaborative learning that can lead to individual understanding of cultural mathematical knowledge.

Tab Welcome

When they enter the opening view of the VMT chat room for Session 1, the group of three students starts with a hesitant discussion of what to do. The situation of an online chat room with a mathematical topic is utterly new to the students and they have to figure out how to proceed from hints given in the initial software interface. They express some indecision, but manage to resolve collectively how to get going.

The VMT interface consists of several panels, which appear on the students' computer screens (see **Figure 4**). On the left is the GeoGebra panel, consisting of several tabs. (This is a multi-user implementation of dynamic geometry for collaboration—the first and so-far only one ever developed and deployed.) In the figure, the "Welcome" Tab is displayed and five other tabs are accessible across the top. On the right is the text-chat panel. It includes a list of the participants currently in the chat room at the top, a list of the recent chat postings, which will eventually scroll off the screen as new postings appear, and, at the bottom, a message-typing area for entering new chat messages to post. In **Figure 4**, the GeoGebra area for constructing dynamic-geometry figures is currently filled with textual instructions from the VMT curriculum designers, intended to guide the student activities. Above this construction area are menus and tool-selection icons that are part of the GeoGebra system. Below is a "Take Control" button to allow one student at a time to interact with the GeoGebra tools and objects. The GeoGebra content is displayed identically to all the students in the team, but only one student at a time can manipulate it. The name of the participant who currently has control is displayed to the right of the button. To the right of that, the GeoGebra tool that is currently selected for use is listed. Below the chat panel, an awareness message lists the names of everyone who is currently typing in the chat message-typing area. Note that the chat panel also includes small squares; these indicate actions in the GeoGebra pane and are color coded to identify the actor. The chat also announces when a participant changes the GeoGebra tab—because looking at another tab does not affect the screens of other team members, but it might be important for the students to know where each other is looking.

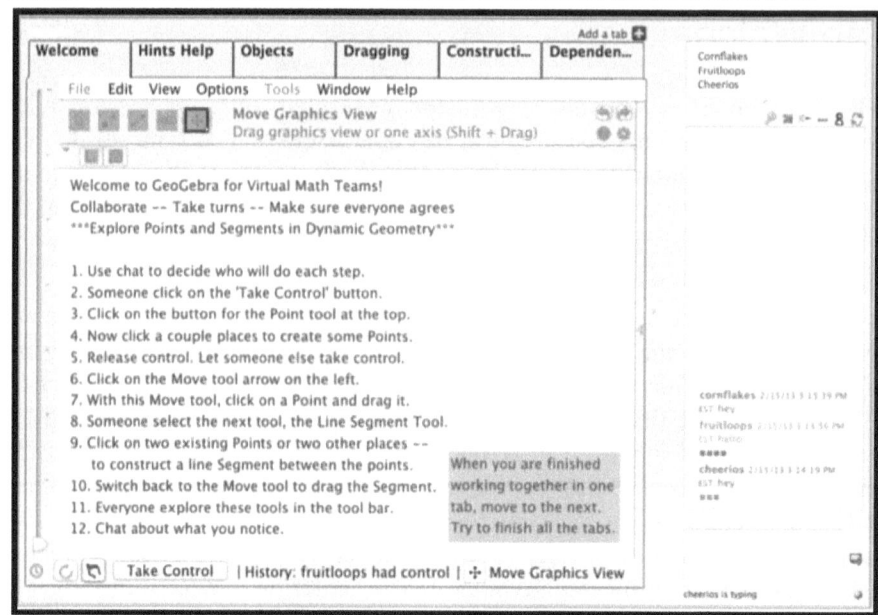

Figure 4. The Welcome Tab of Session 1 for WinterFest 2013.

Figure 4 is a screen image captured from the VMT Replayer, an application that allows a session to be replayed, browsed and studied—usually by a teacher or researcher. The Replayer uses the same technology as the VMT software itself, so that the display is identical to what the students originally all saw on their computer screens. For the sake of conciseness, the chat log excerpts incorporated in this book, like **Log 1**, are filtered to only show chat postings by the students; consequently, they do not include the messages corresponding to actions like starting to type, switching tabs, taking control, dragging points, creating GeoGebra objects or changing tools, although these are available in the dataset.

Log 1. The team meets in the first session.

Line	Post Time	User	Message
3	13:39.4	cornflakes	hey
4	13:57.0	fruitloops	hello
6	14:19.1	cheerios	hey
7	14:45.6	cheerios	whose froot loops
8	14:53.9	cornflakes	xxxxxxxx [name *removed from log for privacy*]
9	15:10.8	cheerios	whose takimg control

10	15:20.1	cheerios	taking*

When the students first enter the VMT environment, they have to make sense of the elements of this interface and figure out what to do and how to do it. The students are presumably familiar with online applications in general and with text chat or instant messaging in particular. They know how to type messages and to start by greeting one another (lines 3, 4 and 6 of **Log 1**) as they see that the others enter the room and their names (anonymous login handles) are displayed. For instance, they know the instant-messaging convention of using a star (*) to indicate a repair to a previous posting (see lines 9 and 10). They are using aliases for reasons of online privacy, so they do not necessarily know to whom the aliases correspond in real life.

Notice that from the start, the students are engaging in dialogic interaction, in which they take turns and respond to each other. In lines 3-6, they begin the chat with the everyday dialog convention of greeting each other. Cornflakes initiates with an informal "hey"—which not only announces her presence to the others, but also elicits a response from them in lines 4 and 6. In lines 7, 9, 21 and 23, Cheerios poses questions. Questions are a common way to elicit responses and thereby start dialog (Zhou, Zemel & Stahl, 2008). Such turn taking has been analyzed by the field of Conversation Analysis as fundamental to verbal discourse (Sacks et al., 1974; Schegloff, 2007). Here, the students adopt it naturally in their text chat. They post greetings, questions and proposals, which elicit responses from others and they then respond to each other's postings. Such dialogical interaction—through responsive turn taking adapted to text chat—is a group practice that they do not have to learn. They bring with them this practice of eliciting and responding to greetings, questions, proposals and the like. Engaging in this familiar form of interaction is *a first stage in the team starting to collaborate*. It establishes the practice of *discursive turn taking* as a recurrent feature in their group process.

> *Group collaboration practice #1: Discursive turn taking.*

Turn taking works differently online than face-to-face (Zemel & Çakir, 2009). In text chat, participants can type simultaneously, rather than waiting for opportunities to take a turn. However, a similar pattern of eliciting and responding to turns takes place, in which text postings are designed to be understood as responses to particular preceding posts—for instance as answers to previous questions. In line 3, Cornflakes displays her greeting to the others. Both Fruitloops and Cheerios must see that posting because they respond to it with their own greetings. Through these greetings, the students each display their active presence in the online interaction environment and their responsiveness to each other. They also display that they are bringing in the greeting conventions from the face-to-face world or the everyday texting world for use in the VMT context.

The three girls in this team know each other as classmates, so the first thing they want to do once they are all present in the chat environment with their online handles is to connect these aliases to their respective personalities. Apparently, Cornflakes knows who is going by the alias "Fruitloops," and sharing that information (lines 7 and **8** of **Log 1**) sorts things out for everyone. This is accomplished by a question/answer elicitation/response pair of postings by different people (Cheerios and Cornflakes). This introduces the use of questions as a driver of interaction.

The next two questions that arise for the group (see **Log 2**) then are:

- "whose takimg control" (line 9)
- "so whoses doing what" (line 21)

Cheerios has taken the lead in asking about identities and raising these two issues in the chat. However, this does not mean the others are passively waiting. Cornflakes has already taken control of GeoGebra twice and Fruitloops has taken control once, before Cheerios asks who will take control. Cornflakes also looks around by changing GeoGebra tabs several times between this question and Cheerios' next question.

Log 2. The team decides to take turns.

9	15:10.8	cheerios	whose takimg control
10	15:20.1	cheerios	taking*
21	16:18.4	cheerios	so whoses doing what
22	16:44.4	fruitloops	who wants to take control?
23	17:30.6	cheerios	xxxxxxxx do you want to [name *removed for privacy*]
24	17:52.2	fruitloops	no... cornflakes you take controll.....
25	18:01.7	fruitloops	who wants to do what steps?
26	18:02.9	cheerios	cornflakes take control
27	18:03.6	cornflakes	no cheerios you can
28	18:14.6	cheerios	cornflakes
29	18:25.4	fruitloops	cornflakes
30	18:33.6	cornflakes	NO
31	18:40.0	cheerios	why not

32	18:52.3	fruitloops	i just took control. lets takes turns
33	19:01.9	cheerios	alright
34	19:03.0	cornflakes	ok

Fruitloops modifies Cheerios' question, "whose takimg control" to be a question of "who wants to take control?" (line 22). Cheerios responds to that by turning the tables and asking Fruitloops if she wants to take control. Fruitloops simply responds "no" and tells Cornflakes to take control. She then immediately moves on to the second issue, formulating it in terms of steps: "who wants to do what steps?" (line 25). Cheerios echoes Fruitloops' statement that Cornflakes should take control, but Cornflakes refuses and tells Cheerios that she can (in line 27, responding to line 24). Cheerios and Fruitloops both insist, by simply repeating Cornflakes' name. To this, Cornflakes shouts back, "NO." In a sense, this is a failed proposal (see Stahl, 2006, Ch. 21) by Fruitloops and Cheerios jointly. However, it does not ultimately fail because their bid at a proposal is not ignored, but is responded to, even if initially in the negative by the other group member. The team pursues this proposal and eventually succeeds in reaching a negotiated agreement.

Fruitloops resolves the mounting conflict in which everyone tries to avoid taking leadership by suggesting that they take turns, in line 32: "lets takes turns." (Note that the instructions in the Welcome Tab begin with the directive: "Collaborate – Take turns – Make sure everyone agrees." Presumably this prompted Fruitloops' recommendation.) The other two students agree to this solution. In this way, the group enacts the collaborative practice recommended by the instructions. While the practice of taking turns in GeoGebra control and action was always present in the interactional environment, it had not been made focal for the student discourse until Fruitloops' proposal bid. Her bid elicits acceptance by the other two by phrasing it as "lets." Whereas previous proposal bids—like those in lines 24 and 26—were declined, this one is accepted. It thereby becomes a successful bid.

In the discussion of control, the first group collaboration practice—discursive turn taking—has already become quite complicated. There have been questions, proposal bids, rejections, failed proposals, counter-proposals, repairs, negotiation, bringing in suggestions and proposing for the group. The general practice of discursive turn taking incorporates many detailed practices that are embedded in the use of human language and in the extensive lexicon, grammar and pragmatics of spoken and written English. The students simultaneously act in accordance with these tacitly understood rules of their normal communication and also, by typing them to each other, display for themselves (and anyone else looking, like us) these usages as legal, acceptable and effective.

By one group member displaying a specific textual action and having it responded to appropriately by another group member—without any objection or breakdown of interaction—the group in effect acknowledges that action/reaction as an acceptable form of interaction. The open-ended set of discursive turn-taking moves is thereby adopted by the group as a group collaboration practice, which is thenceforth available to the group.

The explicit decision to take turns in controlling the GeoGebra system (lines 32-34) establishes a second, distinct, important group practice. Note that the students display their agreement to each other by everyone agreeing with Fruitloops' proposal: "i just took control. lets takes turns." Thereby, they adopt this proposal as a group practice.

In effect, this extends the discursive turn taking from chat to GeoGebra actions, encompassing both these media of online interaction in VMT. The practice of taking turns or responding to each other in the chat stream came rather automatically to the team, but the practice of taking turns with the GeoGebra actions requires considerable negotiation. The team explicitly decides to coordinate the taking of turns with GeoGebra actions through their chat, as was also suggested by step 1 of the instructions. The team will continue to follow this practice throughout their eight sessions together. In adopting this suggestion, the three students start to form themselves into an effective group. This is a second stage in the team learning to collaborate effectively. It introduces a practice of *coordinating activity*—specifically actions involving GeoGebra tools and objects—into the group process.

> *Group collaboration practice #2: Coordinating activity.*

The suggestion to take turns doing GeoGebra actions was given as the first step in the instructions on the screen: "1. Use chat to decide who will do each step." Furthermore, it was implicit in the design decision to include a "Take Control" button in VMT's multi-user interface for GeoGebra. However, the turn taking did not become a group practice until the team *enacted* it. This took place because the team had a breakdown in their ability to proceed when each member declined to take control of GeoGebra to do the assigned tasks—even though two of the members had already voluntarily taken control of GeoGebra to do their own individual exploration. Two of the three students readily engaged in GeoGebra actions individually, under their own agency, but the group had difficulty undertaking any action as a group—by members of the group following a group-action trajectory. Then one person suggested taking turns, putting the advice of the topic instructions into her own words and introducing it into the group interaction as relevant and situated. The other team members immediately and explicitly agreed. As we will see, the team did in fact follow this practice by taking turns on each numbered step of this first topic, as well as with subsequent topics.

It may seem ironic that taking individual turns helps make for a group that works in a unified way. However, this is much like the way that turn taking in conversational discourse builds a social order in which the structured elicitation and response of adjacency pairs results in shared meaning making and intersubjective understanding (Schegloff, 2007; Stahl, 2013c, Ch. 8). Taking turns in the geometry manipulation and construction activities allows everyone to participate in a single group trajectory, where each participant can pay attention to the actions of the others, build on each other's work and thereby display their understanding of the significance of what the group as a whole is accomplishing. If everyone could work in GeoGebra simultaneously, people might tend to focus their attention on their own activities and work individually in parallel, rather than collaborating in shared work. The effect of doing something as an active member of a group is quite different from doing the same action as an individual in terms of the consequences for the shared understanding and joint meaning making of the group. The important thing is not simply being in a group context, but acting as a coordinated and jointly focused team.

In analyzing the group discourse, it is possible to differentiate different roles for the individuals. Cheerios' assertiveness in the chat may reflect reluctance on her part to take control of the GeoGebra tab. The ensuing struggle around this issue poses a problem for their teamwork. It contrasts with the tone of the group later and in their future sessions, where they are usually quick to agree with each other. Even after they decide as a team to take turns, Cheerios does not take a turn at control until they get to step 5, which explicitly says to "let someone else take control." That is about five minutes later (a long time in a chat), around line 61. Even then, Cheerios just drags one point. That is her only GeoGebra action within their work on the Welcome Tab. The three students are not yet acting as a coherent team collaborating; their behavior is still best understood as that of individual actors cooperating.

It is interesting that from their very first acts, the three students seem to adopt different roles in the group discourse. These are not assigned roles, but seem to emerge naturally from the personalities of the individuals and their positioning of each other in the chat interaction:

- Cornflakes is an explorer (early adopter) of the technology; without saying anything, she goes around trying out the available tools. She then guides the others in using unfamiliar tools.

- Cheerios leads the group to action; she tries to get the others to take a next step. She thereby positions the others, and not herself, as the do-ers.
- Fruitloops is the more thoughtful, reflective, questioning and refining member; she re-phrases proposals in a more socially and cognitively productive way.

As we saw at the very start of the session, while Cheerios is asking, "whose froot loops," Cornflakes takes control of GeoGebra and then releases it. Then Fruitloops takes control and releases it and Cornflakes does so again. When Cheerios next asks, "whose takimg control," Cornflakes switches to the other tabs in quick succession, perhaps checking how they relate to the tasks of the first tab. Cheerios next asks, "so whoses doing what," which Fruitloops restates as "who wants to take control?" This is already an example of the students' personal roles. Cornflakes (the technologist) repeatedly controls GeoGebra and also looks around the online environment at the available tabs with their tasks for the future. Meanwhile Cheerios (the social leader) positions herself as group organizer by trying to get someone else to start using GeoGebra. Apparently, Cheerios does not realize that the others have both already started. Her leadership actions may be seen as attempts to cover up the fact that she does not understand what is going on or to gain information about what everyone else is doing. Fruitloops (the reflective one) takes a more reserved role in guiding the group by shifting Cheerios' attempt to find out what others are doing or to get someone else to do the GeoGebra work into an inquiry about who wants to (or is willing to) take control.

The different students' roles persist even after the group coalesces. What is perhaps most interesting about these "individual roles" is how they change as the sessions proceed and how our analysis reveals a complexity to such roles that could be obscured in other analytic approaches. For instance, a traditional research approach—that codes each participant's postings and then counts the number of their postings in each of several categories—might conclude that Cornflakes, who does not post many chat contributions, is a "lurker" who does not contribute much to the group. However, in fact, we will see that Cornflakes often guides the group in learning about how to use the GeoGebra technology for construction. Later, each of the other students adopts the role of technology explorer, following Cornflakes' example and often with her guidance. Similarly, Cheerios' many facilitating comments might indicate that she is a leader, whereas in most of this first session she actually displays the least understanding of what the group is working on—although that is sometimes dramatically reversed in later sessions.

A psychological approach focused on the individual students might interpret Cheerios' feigned leadership as a way for her to avoid the technical work and to cover up her lack of understanding of what is going on. However, in later sessions Cheerios sometimes becomes the most effective explorer of the tools and the most successful problem solver. Others take over her initial role in facilitating the group process and in stimulating reflection.

Perhaps most impressively, Fruitloops' often prescient or instructor-like role in posing questions and raising abstract issues of underlying reasoning is eventually adopted by both Cornflakes and Cheerios by the final sessions.

Through chronologically tracking the longitudinal development of the group discourse, we will see how behaviors that start as individual tendencies gradually become shared group practices. The individual perspectives that are manifested within the team interaction themselves become group practices or features of the collaboration. These shared practices are also influenced by the group's response to the topic instructions. Thereby, community standards, such as mathematical practices and discourse principles modeled in the topic instructions, mediate the group practices and their effects upon the individual and group behaviors.

Both Tomasello (2014, Ch. 4) and Stahl (2006, Ch. 16) stress the central role of personal "perspectives" in collective intentionality or group cognition. According to Tomasello's evolutionary anthropology approach, it was through the understanding of other people's cognitive perspectives that early humans achieved the ability to interact socially in a way that no other animal can. Modern humans have the ability to see the world through someone else's eyes and to recursively recognize that, for instance, the others may know that I know that they know my view. The power of collaboration largely emerges from the sharing of different perspectives (both literally and figuratively) on a shared object.

The ability to see the world from another person's perspective is at work in every discourse action. A discourse utterance is designed by the speaker to be understood by the intended recipients from their perspective. Otherwise, it could not be very effective in eliciting the desired response from them. The recipients not only take the utterance as emanating from the speaker's perspective, but also as designed by the speaker for them, as understood by the speaker. The recipients respond with an utterance that is similarly (symmetrically) designed for its recipients. The structure of the adjacency pair incorporates the recursive understanding of other people's perspectives. In fact, most utterances in an ongoing interaction are both responsive to prior acts and eliciting of future responses. Thus, an elicitation-and-response pair of utterances both projects and confirms a shared understanding in its interactionally constituted meaning. It bridges across the differing perspectives of speaker and hearer on their shared world to construct intersubjective meaning.

There may be some physiological basis for perspective sharing in the brain's mirror neurons (Gallese & Lakoff, 2005), which are far more highly developed in humans than in other primates. However, shared perspectivity is primarily the result of more recent cultural evolution (Donald, 2001), especially in the transition to agriculture and the establishment of village life (Seddon, 2014). Because it is cultural rather than genetic, it is a skill that must be developed again in the enculturation of each human child. This takes place interactionally and we will see examples of it in the team's exchanges of mathematical viewpoints.

It may be productive to consider the students' individual personal differences—including their different perspectives on the shared world—to be related to the individual team members being at different "zones of proximal development" or ZpD. This is a concept proposed by Vygotsky (1930) to indicate what a person is developmentally ready to learn, especially when guided through interaction with others. We might say that in these VMT chats, Cheerios initially seems ready and eager to learn social skills of collaboration; Cornflakes seems to be repeatedly engaged in trying to learn technical skills of working in GeoGebra; and Fruitloops seems to be oriented to learning theoretical skills like reflecting on why something is true. In the early sessions, each of the students seems to lead the group in the direction of their own apparent ZpD. Increasingly over the sessions they each appropriate the other students' approaches, converting them into shared group practices. The students' success on the session topics derives from the interaction of their individual contributions, merged into a unified process of shared meaning making.

However, we are less interested in focusing exclusively on the individual as learner—which has been so intensively studied by many others in the past and present—than in observing them in small groups. Even early in the first session, the students start to constitute themselves as a group. They begin to refer to themselves collectively in the first person plural, as "we" (lines 36, 38, 40, 44) or "us" (line 40). They are discussing what they should do as a group. A set of students meeting inside of a physical or virtual environment is not necessarily a collaborative group. The students must act as a coherent collectivity that works together as a unity and that begins to refer to itself that way. One way they start to do this is to talk about themselves as a collective subject (first person plural) in their chat (Lerner, 1993). This initiates a third collaborative practice, *verbally constituting their group as a collective unity*.

> *Group collaboration practice #3: Constituting a collectivity.*

Having resolved their first question, the students turn to their second question: the question of agency—both individual agency (what *I* want to do) and group agency (what *we* should do). This is the universal existential question, which every individual and every group must pose in some way at each moment: What should we do here and now? Most of the time, people confront this question while caught up in the midst of an activity trajectory, with a history of commitments, motivations, resources and decisions—and moving toward some complex of goals and projects. They have many shared practices that are procedures for going on. However, as they begin their first session together, this virtual math team has little shared history, agreed-upon aims or established practices to steer their action. They find themselves in a context with some general structure as participants in an after-school activity involving geometry. While that is not enough to specify what exactly they should be doing, it is enough to orient them to the text in the VMT environment. This text begins with the instructions:

1. Use chat to decide who will do each step.

2. Someone click on the 'Take Control' button.

This sufficed to motivate the group's first two concerns: who should take control and what should they do?

In earlier iterations of the VMT Project, it became apparent that users needed some form of instruction or guidance in the use of the VMT technology and the GeoGebra tools, as well as in best practices for working together online on mathematical tasks. In one trial, even groups of graduate students well experienced in computer technology found it difficult to get started without some kind of manual or training (Stahl, 2013c, Ch. 9). Tutorial texts were then produced for the next iterations of VMT, but it was clear that users did not study them.

In WinterFest 2013 and in the preceding teacher professional training for it, the VMT environment designers tried to provide the needed guidance in the form of instructions inserted in the GeoGebra tabs of the chat rooms. We have already seen in the discourse of **Log 2** that the Cereal Team was guided by the displayed text in the Welcome Tab shown in **Figure 4**. However, they had to take up the various suggested practices themselves, discuss them, make sense of them in their current context, negotiate how to implement them and agree to follow them. This is what is meant by the group "*enacting*" the suggestions: Social order, practices, tools and organizational structures are not simply given to people, but must be enacted by them through discourse (Carreira et al., 2016; Latour, 2007; LeBaron, 2002; Overdijk et al., 2014; Rabardel & Bourmaud, 2003; Weick, 1988; Zemel & Koschmann, 2013). In this book, we want to see how the team enacts the embedded instruction in collaborative dynamic geometry, including suggestions for effective collaboration. We see this continue to unfold in the Cereal Team's first session.

Having decided to take turns clicking on the 'Take Control' button, the students use chat to decide who will take a turn (see lines 41, 45, 46 in **Log 3**). They note that the interface tells them who is in control, or that no one is (line 35). They also note that the displayed text includes numbered steps, which they decide to follow (line 38—note that Fruitloops already referred to "steps" in line 25).

Log 3. The team expresses their confusion.

35	19:26.6	cheerios	it says no one has control
36	19:30.1	fruitloops	what do we do know?
37	19:44.3	cheerios	i am not sure cornflakes do u know
38	20:17.4	cheerios	i think we have to follow the numbered list
39	20:20.0	cornflakes	uh mo
40	20:38.4	cheerios	so l ets do that and we will figure it out as we go
41	20:50.6	fruitloops	someone else take control for now
42	20:53.1	cheerios	lets*
43	21:17.5	fruitloops	just follow the welcome thing
44	21:32.9	cheerios	yeah so we are on #3
45	21:56.2	cornflakes	ok someoen else take control
46	22:05.3	fruitloops	someone take control
47	22:13.1	cheerios	whats happening?
48	22:18.7	fruitloops	idk
49	22:30.8	cheerios	i am so lost
50	22:36.6	fruitloops	i took control. what should i do?
51	22:42.3	cornflakes	make a line
52	22:50.9	cheerios	i am not sure #3 i guess?
53	23:10.4	cornflakes	no i already did 3, do 5

This excerpt reflects a "breakdown" in their interaction. Cheerios, who tries to lead the group process, says she is not sure what to do and that she is "so lost" (line 49). Fruitloops also says "idk" (line 48)—*I do not know* what is happening. Cornflakes is doing some work, but not collaboratively with the others. Heidegger (1927) has argued that breakdowns in the smooth functioning of people taking action in the world can serve to reveal existential structures that are normally tacit and hard to observe (Koschmann, Kuutti & Hickman, 1998;

Stahl, 1993). In this case, we have a breakdown in group collaborative action rather than in individual cognition, which reveals a necessary group practice that is missing in **Log 3**, but will already be present by the postings in the next excerpt.

The team overcomes the breakdown in its work by taking up the guidance offered in the instructions, to follow the numbered steps. From an observer's perspective, this is an obvious thing for the team to do since the instructions tell them it should be done. Furthermore, each of the students displays some sense of how to do it (Cheerios in line 38, Fruitloops in line 43, Cornflakes in line 53). Although the individual students already follow the instruction to some extent, the team as a whole has to adopt the procedure of following the numbered steps as a collaborative procedure for it to be effective in supporting the team effort. This is similar to the case explored in (Stahl, 2006, Ch. 13), where a face-to-face team of students had to enact—as a group—a new shared understanding of the significance of a meaningful ordering of items, although it was obviously intended by the designer of the software environment they were using.

In **Log 3**, the group verbally agrees to follow the steps in the instructions (lines 38-43). Then they begin to actually follow them: Cheerios suggests that they are at step 3 (lines 44 and 52), but then Cornflakes corrects this to step 5 (line 53). Following the instruction's numbered steps establishes a group practice that allows the group to proceed in their collaboration and their mathematical work in a way that keeps going through an explicit, numbered system of *sequentiality*. This is a fourth group collaboration practice adopted by the team.

> *Group collaboration practice #4: Sequentiality.*

Sequentiality will prove important for allowing the group to engage in long, intense discussions (hour-long chats) and complicated mathematical tasks (geometric constructions involving multiple sequenced actions). Both knowledge-building discourse and mathematical problem solving require the ability to stay focused and to continue for long periods. While informal conversation can consist of brief interchanges, achieving the goals set in the VMT sessions requires long sequences of interaction (Stahl, 2011a). The group's adoption of the sequentiality practice here permits them to work effectively on the VMT goals in their future sessions.

As we see in **Log 3**, the practice of following the sequentiality of the numbered steps does not, however, work smoothly from the start. While Cheerios and Fruitloops are talking about what the group should do in lines 40 to 44, Cornflakes proceeds to take control of GeoGebra and actually constructs a point A and a line segment AB. When she is finished, Cornflakes says "ok someoen else take control" (line 45). The "ok" signifies that she has accomplished something and the rest of her post requests the others to build on what she has done. However, in lines 46 to 50 Fruitloops and Cheerios indicate that they do not know what is going on, that is, what is the action trajectory that they should be taking further. They probably saw the GeoGebra points and segment appear on their screen, but did not know where they came from. In typical fashion, Cornflakes has gone off individually to explore the technology and to complete the construction specified in the topic steps 2, 3, 4, 8 and 9. However, she has not involved other people, announced in the chat what she was planning to do or described in the chat what she did. She has not yet adopted effective collaboration practices.

Although the points and line segment that Cornflakes created in GeoGebra should have appeared on the computer screens of Fruitloops and Cheerios as well, it is not clear that her teammates saw them or understood that Cornflakes had constructed them. Seeing the appearance of these geometric objects *as* results of someone's construction actions is something that has to be learned—Goodwin (1994) calls this "professional vision" and Wittgenstein (1953) calls it "seeing as." The students have to learn to see the appearance and movement of objects in the GeoGebra space *as* specific intentional actions of the group member who currently has control of the GeoGebra tools. Learning this skilled vision is facilitated by collaborative communication in which, for instance, the person in control states what she will do next or what action she has just taken. This indicates where everyone should look and thereby contributes to the joint attention to a common object as understood with a shared meaning, which is a hallmark of collaboration

The group's attempt to communicate about what they are doing is confused in **Log 3**. Fruitloops takes control, but then does not know how to proceed. Cornflakes types, "make a line" (line 51), suggesting that Fruitloops also experience constructing a line. When Cheerios tries to orient the team to what they are doing by reference to the numbered list, she proposes that they should be doing step 3. However, Cornflakes responds "no i already did 3, do 5" (line 53). So their adoption of their new group practice of sequentiality as following the instruction's steps is not working smoothly yet.

Because they are working online, it is at first hard for them to tell what the others are doing—such as that Cornflakes constructed a point or a segment. They need to develop ways of informing each other as they work. The team is sometimes confused about what to do in this strange environment. However, they persevere by chatting with each other. They address each other—starting with their initial greetings—and instruct each other by responding to questions and proposals, and eventually by assenting to agreement. In addition, they have been told by their teacher to be "descriptive" and to state what they are doing in the chat, and they will remind each other of that periodically during their sessions (e.g., lines 188 and 201 later in this session). Gradually, they start to follow the steps outlined in the first screen. They coordinate their actions by sharing with each other what numbered step of the instructions—e.g., step 3 (line 44) or step 5 (line 53)—they should all attend to together.

We can take this as a further stage in the team learning to collaborate. A second breakdown in their work together occurs as they realize that they do not know what each other is doing. In particular, Cornflakes has gone off and completed a number of tasks, but the others do not know what she has done and therefore they do not know what they should do to continue what she has started. The team members start to chat about this problem and they begin to refer to their actions by the numbering of the steps in the instructions. So the team begins to adopt some practices of chatting about what they have done and incorporating the step number in their descriptions. Referencing the steps in their discourse leads to the practice of following the numbered steps in their actions. This coordinates things so that the whole group is present at the same step of the instructions. At this stage, the group has developed a form of *co-presence*, in which everyone is aware of the group action trajectory and present at the same step.

> *Group collaboration practice #5: Co-presence.*

From then on, the team orients to the numbered steps in the instructions on the screen. This has the positive result of getting the team out of their breakdown situation and allowing them to move on through the prescribed steps. It provides a procedure for the group to accomplish the assigned task by working together. As we shall see, coordinating their discourse and geometry action allows the team to co-experience an intersubjectively meaningful world and build shared understanding. As the team goes through the steps collectively, each member of the group then tries out each GeoGebra action for themselves—either vicariously by observing a peer doing it or personally by doing it herself.

While the sequentiality practice facilitates co-presence, at the same time it has some disadvantages in that the steps on the screen may not be optimally structured, and in that concentration on the sequence of steps may distract from reflection on the mathematical goals of the topic. It is likely to lead to the typical student orientation to completing assigned action procedures without thinking about their educational significance. What the team is co-present at is the numbered instruction, rather than the unfolding mathematical task that should emerge from the work of following the instructions. The team may be oriented more toward the numbered tasks given in the instructions than to the intended geometric content: for instance, they might say they are doing step 9 rather than that they are constructing segment AB.

In fact, the numbered steps as given in some of the instructions do not even always correspond well to meaningful action turns for participants. For instance, by constructing point A and segment AB, Cornflakes has actually completed steps 3, 4, 8 and 9. The numbered steps are not each whole, separable tasks. For instance, steps 2-5 go together to construct a point, steps 5-7 are for dragging the point and steps 8-10 are for constructing and dragging a segment. When Cornflakes tells Fruitloops to do step 5, Fruitloops takes control and selects the Move tool, but then wonders "okay now what?" (line 54 in **Log 4**). She then goes on to drag points A and B of the segment, which is really step 7 or step 10. Cornflakes sees the points move and says, "good."

Log 4. The team constructs a shape.

54	23:10.5	fruitloops	okay now what?
55	23:43.6	cornflakes	good
56	24:02.8	cheerios	it says to release control
57	24:12.8	cheerios	and then do #6
58	24:15.7	cornflakes	then release control
59	24:17.7	fruitloops	now someone erlse continue
60	24:24.7	fruitloops	released
61	24:29.0	cornflakes	cheerios will
62	24:44.4	fruitloops	take control and explore with the other toolos
63	24:57.3	cheerios	i just did 6
64	25:17.5	cornflakes	ill do 7 then
65	25:21.6	cheerios	ok
66	25:26.1	fruitloops	ok
67	25:50.2	cornflakes	ok done
68	25:51.0	fruitloops	do likie 9 and 10 also
69	26:24.1	cheerios	what about 8
70	26:41.1	fruitloops	yeah
71	26:46.6	cornflakes	there
72	26:51.1	fruitloops	can i go next?
73	26:57.1	cheerios	yes

74	26:57.3	cornflakes	yeah go ahead
75	27:14.1	fruitloops	so we just play around with it?
76	27:19.9	cheerios	now the triangle is bigger
77	27:22.0	cornflakes	i guess pretty much
78	27:35.5	cheerios	are we on 11
79	27:41.0	cornflakes	yes mam

They tell Cheerios to take control (lines 61 and 62). She does, and she drags point A. She reports that she did step 6 and Cornflakes volunteers to do step 7. She likewise drags point A. Now all three participants have dragged point A— each one more vigorously than the previous one. Cheerios is still confused about what steps are done when. Furthermore, no one has dragged the segment as a whole (step 10). Fruitloops requests control and inquires "so we just play around with it?" (line 75), to which Cornflakes responds "i guess pretty much." Fruitloops playfully adds a number of connected segments.

First, Fruitloops drags the vertices of the triangle that Cornflakes had created. Then she creates a new point D. She also constructs a point E on one of the sides of the triangle. She drags point E and sees that it remains on segment BA, before she drags it to the end of the segment. Fruitloops now asks, "how do we get the line to connect to the piont?" (line 80, **Log 5**). Before anyone can answer—just five seconds after posting her question—Fruitloops selects the GeoGebra segment tool and connects her new point D to her point E, which is very close to point B of Cornflakes' triangle. Fruitloops' GeoGebra constructions and her question display a growing sense of how points and lines are "connected" in dynamic geometry. They are not just visually connected, but a point can be constructed on a line and be confined to that line during dragging. Also, an existing point can be used as an endpoint for a new line segment. This is all displayed in Fruitloops' playful exploration of the use of GeoGebra tools—see (Çakir & Stahl, 2013; Çakir, Zemel & Stahl, 2009) for examples of how graphical actions in VMT can display ones mathematical understandings to teammates.

Log 5. The team connects objects.

80	27:44.5	fruitloops	how do we get the line to connect to the piont?
81	27:45.2	cheerios	kk
82	28:00.0	fruitloops	nevermind

The students learn to see what each other is doing by:

- Taking turns each doing the same GeoGebra actions, so that they experience the use of the tools first-hand and can see the results as similar to appearances when their teammates were in control.

- Discussing what they are doing in the chat and guiding each other through the chat to do the same things in the geometry.

From this, they start to follow each other's GeoGebra actions and become co-present at the same objects. The actions then become visibly meaningful. Just as they can communicate through words in the chat, they start to be able to communicate with each other through observable, interpretable actions in the GeoGebra workspace.

Understanding dynamic geometry involves an integration of manipulating geometric objects spatially and reasoning about them verbally. This is true for both group understanding and individual understanding. Geometric phenomena, relationships and arguments have to work both conceptually and in constructions. One can talk about the equality of sides of an equilateral triangle using mathematical terminology; one can view the lines and letters on the computer screen; one can measure the side lengths on a drawing; one can think about an ideal triangle whose lines have no width. This complementarity is built into the nature of geometry: its implementations in language, drawings, software and imagination. It must be reflected in team and student learning by a unity of spatial manipulation and mathematical discourse.

The team members coordinate GeoGebra actions with chat in order to explore and demonstrate for the team (both for themselves and the others) features of the dynamic-geometry system. The team learns to observe what the person in control is doing in GeoGebra and they chat about it just enough to share their understanding of what is taking place, at least at a superficial level of which numbered step is being carried out. In this stage, they develop a meaningful form of *joint attention* to the GeoGebra actions, with mechanisms for coordinating and maintaining that group attention.

This leads to each member of the team then trying out the GeoGebra actions for themselves. This is another stage in the team learning to collaborate. They often repeat what each other has done. This gives each person a hands-on experience with the particular kind of action. It also displays to the group that they each understand what the others have done and how to do the action themselves. They are engaging in joint attention by all team members to a current object of team concern. At this stage, it may be necessary for each student to take the same action in order to develop the same understanding. Doing so will enable each student later to learn from what another student does without repeating it, because they will have learned to share each other's perspective. The individual perspectives merge into a group perspective, and the visual and action-oriented attention to geometric objects becomes a joint attention by the whole team to those objects as understood as having the same shared meaning.

> *Group collaboration practice #6: Joint attention.*

By identifying adoption of a group practice, like joint attention, as a particular stage in learning, we do not mean to imply that the team completely mastered that stage. As we will see, their mastery is generally partial and fragile; they often fall back on previous forms of floundering.

Even if everyone is oriented toward the actions of the person who has control of GeoGebra activity, they do not yet necessarily fully follow what is being done or its implications for understanding dynamic mathematics. Many actions that the students engage in using GeoGebra are hard to follow or to infer a guiding purpose. There are multiple ways to accomplish basic aims and observed actions can be understood as attempts at different goals—or as not having a clear purpose. So joint attention to changes in the GeoGebra display does not necessarily imply joint attention to shared meaning making.

Learning how to use a dynamic-geometry software system requires considerable exploration and trial. For instance, a line segment can be constructed using existing points or by creating new points in the process of defining the segment. By using existing points, one can attach a new segment to an existing segment, forming a figure that can be dragged around in complex ways and remain connected. Before posting line 80 about connecting a new line to an existing point, Fruitloops, perhaps accidentally, placed a new point E on an existing segment. In dragging it, she could see that the point remained on the segment. No one remarked on this. It could have been an important discovery and a shared understanding if the team had discussed it—but they did not.

Connecting one segment to another through a shared point and constraining a point to stay on a segment were the first forms of dynamic-geometry *dependency* that the team encountered. It would take many more such encounters for the team to become aware of the significance of this and to be able to articulate such dependencies. Such knowledge comes gradually, as one explores. The goal of the topic instructions is to kick-start such exploration and then to keep things open enough to allow for free exploration ("playing around") and serendipity. We will see in the later topics that the team starts to recognize the significance of different construction moves and to adopt corresponding group construction practices after the team has developed a basis for effective collaboration in this first session.

The team has now constructed a number of points and joined them together with connected segments to form "a very interestiong shape" (see

Log 6, line 85). The team's construction is shown in **Figure 5** (which shows the VMT chat room displayed within the VMT Replayer; across the bottom of the Replayer screen are controls for browsing through the session).

Log 6. The team describes its shape.

83	28:12.7	cheerios	what now?
84	28:27.9	cornflakes	chat about whatr we njotice?
85	28:41.6	cheerios	well its a very interestiong shape
86	29:04.0	cheerios	a rectangle and a triangle thats mushed together
87	29:10.8	cornflakes	its like a polygon
88	29:14.8	cornflakes	right?
89	29:38.5	cornflakes	no curved edges cause its made of a line segment and line segments are lines and lines that dont have curves
90	29:22.5	cheerios	it has 6 sides
91	29:40.1	cheerios	and obtuse and acute angles no right angles
92	29:34.6	fruitloops	how do i make it smaller?
93	29:57.4	cornflakes	yuppies no right angles
94	29:58.9	fruitloops	should we move on?
95	30:03.5	cornflakes	yessiree
97	30:11.3	cheerios	i think we should
99	30:16.0	fruitloops	Okay lets go

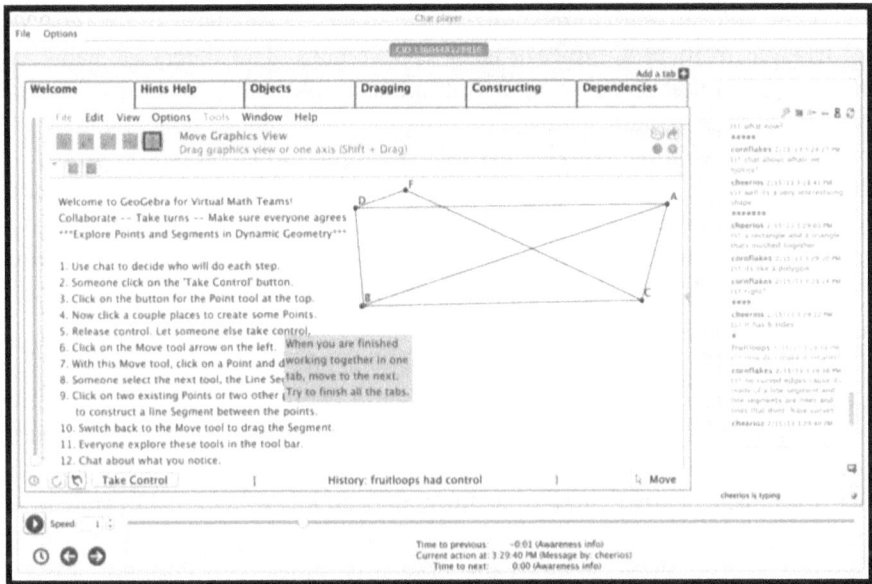

Figure 5. The team constructs a polygon.

The students do not follow the instruction step 7 to drag the new points or step 10 to drag the segments. They are selective about which tasks they choose to do. When they chat about what they notice—in response to step 12—they discuss features of the overall shape as a fixed figure, not as a dynamic-geometry figure. They describe it in terms of its visual appearance as "a rectangle and a triangle thats mushed together" (line 86), rather than as a dynamic construction of connected segments. They then all decide to move on to the next tab for the session.

The description in line 86 is a mathematically unsophisticated way to describe what they constructed, not only in terms of the wording (e.g., "mushed together"), but also as a combination of visual shapes, rather than as relations among geometric objects. Identifying common visual shapes is considered the first of the "van Hiele levels" (van Hiele, 1986; 1999) (see deVilliers, 2003, p. 11). This is a theorized series of increasingly sophisticated levels of geometric reasoning, which students typically progress through as they learn to engage in justification and ultimately axiomatic proof. The levels include:

a) Visual recognition (as in line 86),

b) Analysis of geometric properties,

c) Logical ordering and

d) Deductive reasoning.

Young children learn to recognize basic shapes like square, circle, rectangle and triangle from prototypical visual appearances. The theory of van Hiele levels suggests that students have to progress through the successive levels in order to engage at later levels. For instance, they cannot understand formal proofs without first recognizing the relationships among geometric properties. The ability to identify dependencies and to design construction protocols for building geometric figures with dependencies may be best taught by facilitating the successive movement of students through something like the van Hiele levels to an understanding of the kind of cognition associated with deduction, as in proofs (Stahl, 2013c, Ch. 9).

As van Hiele (1999) recommends, "instruction intended to foster development from one level to the next should include sequences of activities, beginning with an exploratory phase, gradually building concepts and related language, and culminating in summary activities that help students integrate what they have learned into what they already know" (p. 311). That is the approach of the VMT curriculum. In this book, we will see—primarily in Session 3—how the Cereal Team gradually moves beyond the level of observing common shapes and how well the sequence of topics in the curriculum effectively provides the kind of guidance that van Hiele suggests.

Beyond asking the students to describe what they notice in the created figures, the task in the Welcome Tab was designed to provide a first experience with the dynamic character of points and segments as movable or "drag-able." The students did not fully realize this intention. For instance, if they had dragged point F of their figure they would have changed the outer shape, and dragging other points would alter the size of its angles. The students could then have noticed that the visual shapes change dynamically, but that certain relationships, such as certain segments staying connected, are maintained. By constructing a figure but not dragging its points, the students have succeeded in using some basic GeoGebra construction tools, but they have largely missed the intention of the introduction to the dynamic character of the objects created.

The team moves on to the next GeoGebra tab. Fruitloops proposes: "should we move on?" (line 94). The first tab included a small note that said, "When you are finished working together in one tab, move to the next. Try to finish all tabs." Fruitloops enacts this instruction with her question. By inquiring in the chat if everyone is ready to move on, Fruitloops initiates a group practice for *closing a topic*.

As is common in conversation, the discussion of a particular topic is opened and closed by interactional practices or conversational methods (see Schegloff & Sacks, 1973). The team came to this session already knowing how to open an online session by logging in and saying hello in the chat. At this point, Fruitloops enacts a practice for closing work on a GeoGebra tab. She inquires in the chat if the team is ready to move to the next tab. Given that their chat room for this session contains six GeoGebra tabs and that they had already spent more than a quarter of their hour on the first one, time pressure must have been a consideration. No one objects in response to her question, and they each change to the next tab. The success of Fruitloops' method of closing the topic serves to establish this approach as a group practice to be followed in the future.

> *Group collaboration practice #7: Opening and closing topics.*

Defining old and new topics by opening and closing discussions of them is an important technique for extending sequentiality. Short sequences of discourse are defined by adjacency pairs or responses to elicitations. The minimal pair of two utterances (spoken or posted) can be elaborated by secondary utterances which introduce them, wrap them up or insert embedded sequences, such as answering a question with another question or delaying a response pending a repair or clarification (Schegloff, 2007). These short sequences can be grouped into longer sequences that are relevant to each other in that they all discuss a current topic. Boundaries are formed between a current topic and other topics, as it is closed and the next topic is opened. This hierarchical structure of utterance, response pair, extended and embedded pairs, longer sequences, topics and series of related topics is necessary for complex discussions, including those involving mathematical problem solving and collaborative knowledge building (Stahl, 2011b). Meaning is established in the response pairs and then modified as it is incorporated into larger argumentative structures. This facilitates understanding of meaning by the group and its members as they make meaning situated in the context of continuing interaction.

As we saw in lines 94-99, the closing of discussion of the topic concerning the Welcome Tab involved an elicitation of agreement by Fruitloops, responses from Cornflakes and Cheerios and then conclusion by Fruitloops. The sequencing of topics in an extended discourse like Session 1 defines a flow of time, which can be experienced and referenced by the participants (Sarmiento & Stahl, 2008a). As we will see throughout the sessions of the Cereal Group, the students refer back to previous shared experiences associated with topics of discussion, and they also project future topics for investigation. The practices of building continuing sequentiality, long sequences and successive topics constitute an interpersonal temporality, which is shared by the group. The establishment of this subjective but shared form of time progressions is an organizing practice that the team adopts for their work together, even as they move on from their first tab.

> Group collaboration practice #8:
> Interpersonal temporality.

Tab Hints Help

The next tab, the Hints Help Tab (**Figure 6**), is just intended to provide advice about how to adjust the computer image for optimal viewing, depending on the resolution of ones computer. This tab is not intended for collaborative work. Most of the actions discussed in this tab—such as zooming and shifting the image around with the Move-Graphic tool are single-user commands and do not affect the views on other people's computers.

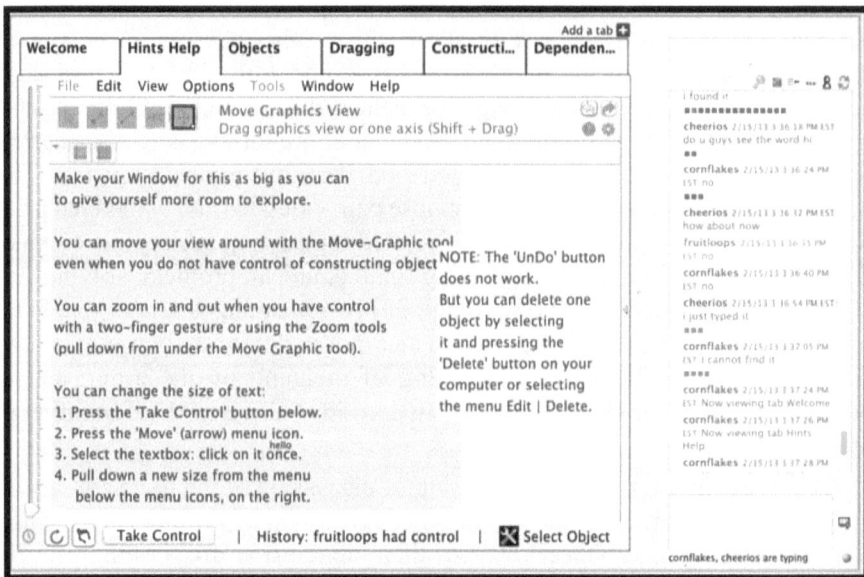

Figure 6. The Hints Help Tab.

This tab caused considerable confusion for the group members because they could not see most of what each other did. Either they were doing actions that did not affect each other's computers or they constructed text boxes that were out of view due to different zooming of their screens.

Cornflakes moves the text boxes of this tab around and the others see the result, although they do not understand how she accomplished that. Then Cheerios succeeds in creating a new text box with the word, "hi" in it (**Log 7**, line 128). However, it is off screen for the others. Fruitloops also creates a new text box with the word "hello" in it, but the others do not notice it. (It appears between points 2 and 3 in **Figure 6**).

Log 7. The team cannot see each other's work.

128	36:18.8	cheerios	do u guys see the word hi
129	36:24.1	cornflakes	no
130	36:32.5	cheerios	how about now
131	36:36.0	fruitloops	no
132	36:40.3	cornflakes	no
133	36:54.6	cheerios	i just typed it
134	37:05.0	cornflakes	i cannot find it
143	37:34.0	cheerios	frootloops do u see it
147	37:46.7	cornflakes	i cant find it
148	37:58.7	cheerios	its a box and says ABC
149	38:02.2	fruitloops	do you see my hello?
150	38:06.3	cornflakes	no
151	38:07.6	cheerios	noo
152	38:17.8	fruitloops	i dont understand it
153	38:27.9	fruitloops	would you like to move on to objects?

Here we see by its absence the importance of *shared understanding* for group collaboration.

> *Group collaboration practice #9: Shared understanding.*

One student creates something on the shared interface, but the others do not recognize it. The students are actively seeking evidence of shared understanding in **Log 7**. However, it eludes them. They direct the attention of the others and ask if they see what has been created. They state that they fail to see what should be a focus of group attention. They try repeatedly to fix the problem and they express their frustration at not being able to establish shared understanding here.

Shared understanding is closely related to practices of joint attention and co-presence. In the excerpt of This tab caused considerable confusion for the group members because they could not see most of what each other did. Either they were doing actions that did not affect each other's computers or they constructed text boxes that were out of view due to different zooming of their screens.

Cornflakes moves the text boxes of this tab around and the others see the result, although they do not understand how she accomplished that. Then Cheerios succeeds in creating a new text box with the word, "hi" in it (**Log 7**, line 128). However, it is off screen for the others. Fruitloops also creates a new text box with the word "hello" in it, but the others do not notice it. (It appears between points 2 and 3 in **Figure 6**).

Log 7, we can see that the students are co-present, involved together with the same tasks. However, they cannot attend jointly to the same textual objects on the screen because they are zoomed off the screens of some of the students or are too small to be noticed. The students work hard to repair the problem. They try to direct attention through chat questions and descriptions. They announce what they are doing and check whether the others observe it (e.g., lines 130, 133 and 148). During breakdowns in shared understanding, practices of *repair* are common (Schegloff, 2007). There are established practices in language for both the people who have created something that is not seen or understood and for the others. The students naturally enact some of these repair practices as they work to establish shared understanding.

> *Group collaboration practice #10: Repair of understanding problems.*

This tab was intended for individual usage. So the collaborative practices of the team were not very effective in working on this tab. In fact, the important condition of shared understanding could not be established in this situation. Given the lack of this essential condition for collaboration, the group cannot work together on this tab. The team consequently decides to move on.

An implication we can draw from this episode for VMT project design is that in the future the information from this tab should be presented to students individually, not in a group collaboration context.

We move on with the team to their work on the next tab.

Tab Objects

The next tab (Objects, **Figure 7**) is intended to provide further experience with the dynamic character of points, segments and circles in dynamic geometry. It provides several example figures (on the left of the textual instructions) and asks the group to drag specific points to explore the constrained movements that are possible.

Rather than following the instructions in detail, the team experiments with the GeoGebra interface, figuring out how to display a grid across the tab or how to change the color of a point. Perhaps they are influenced by the previous tab—which they did not succeed in exploring—to investigate the multi-user GeoGebra interface features more. They are confused by details of the software features, such as that the display of the grid is not shared in everyone's view. They also have trouble with the text box being shared—perhaps because of technical software issues, like that they do not close their text entry box or locate their text box where it will be more visible to others. By the end of their work in this tab, the team succeeds in constructing the polygons and circles shown to the right of the instructions text box.

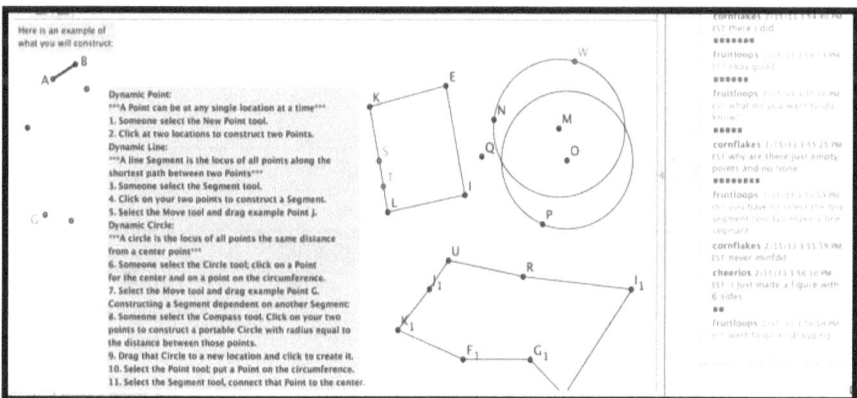

Figure 7. The Objects Tab.

Fruitloops starts the work in this tab by inviting Cornflakes to take her turn at leading in this tab. Cheerios restates the recommendation of their teacher that they explain in the chat what they are doing in GeoGebra (**Log 8**, line 188), and that they describe for each other what they are trying and what they are observing (line 201). This may be an attempt to avoid the earlier failure to establish shared understanding. It corresponds to the team's group practice of reporting in the chat what they have done in GeoGebra, as part of coordinating their activity. Cheerios also asks the others to try to change the colors of points, although she herself is still not taking control and undertaking actions in GeoGebra.

Log 8. The team explores simple figures.

186	43:37.3	fruitloops	flakes explore

187	43:52.3	cornflakes	just made a line segment
188	44:04.1	cheerios	explain what you are doing too
189	44:15.6	fruitloops	can you change the color of it?
198	44:48.4	cheerios	can u make it a different color
199	44:51.8	cornflakes	change the color
200	45:18.2	fruitloops	i made two cirlces
201	45:23.1	cheerios	be descriptive guys
202	45:32.3	fruitloops	okay
203	45:41.0	cornflakes	ok so 2 black circles
204	45:42.1	fruitloops	i made a couple of piotns
205	46:28.0	fruitloops	i made two pionts on line kl
206	46:30.2	cheerios	can you make it colorful
207	46:30.7	cornflakes	ok so we have a poly gon with points k,e,s,t,l,i
208	46:33.3	fruitloops	KL*
209	46:50.4	fruitloops	quadrilateral
210	46:58.9	cornflakes	yeah there we go
211	47:01.2	cheerios	isnt it a rectangle
212	47:02.8	cornflakes	a quadrilatreral is what>?
213	47:14.7	cornflakes	no its not syymmetrical
214	47:21.7	cornflakes	its a quadrilateral
215	47:23.0	cheerios	oh i see
216	47:24.3	fruitloops	quadrilateral- a four sided shape?
217	47:28.4	cornflakes	yess
218	47:35.1	fruitloops	someonee else explore
219	47:39.0	cheerios	i will
220	48:00.3	fruitloops	okay, walk us through what you're doing
221	49:18.1	cornflakes	walk us through ehat your doing
222	49:26.9	cheerios	i made either a complementry or a supplementry angle but iam not sure

Note that both Fruitloops and Cornflakes respond to Cheerios' request for descriptions in chat of what everyone is doing in GeoGebra. Cornflakes' post in line 203, "ok so 2 black circles," does multiple work. She responds affirmatively to Cheerios' request with the "ok" and goes on to not only acknowledge Fruitcakes' description in line 200, but to further specify it in terms of the circle's color—a topical feature that the group was focusing its joint attention on.

Then in line 205, Fruitloops repairs her own utterance from line 204 by saying, "i made two pionts on line kl." Here, she clarifies "a couple" to mean "two." But, in addition, she points to the location of the points as being "on line kl." In line 208, she even repairs this: "KL*," using the convention from texting of indicating a repair with an asterisk. By historical tradition, points in GeoGebra are supposed to be indicated by capital letters. In addition, the lower case "l" might be mistaken for a one or an "i." Fruitloops' repairs are carefully formulated to point unambiguously to the GeoGebra object she has created in a way that is designed to be understood clearly by the recipients of the chat.

Fruitloops here introduces the use of the letters labeling points in GeoGebra as a means of pointing to them in the chat. In linguistics, pointing at something is called "deixis." Pointing is a universal practice among humans in face-to-face settings—generally the first intersubjective gesture that an infant learns (see Vygotsky, 1930, p. 56). In an online setting, pointing can become problematic. One solution is to describe an intended object textually. However, that can be cumbersome and is not always effective.

The use of letters for indicating points was invented by the early Greek geometers (Netz, 1999). It has proven to be an effective way to index or indicate a geometric point, even in complicated constructions. According to linguistics, discussions use indexical devices—like names, labels or special words such as "this" and "that"—to build up an "indexical ground of deictic reference" (Hanks, 1992), a conceptual space in which people participating in the discussion can keep track of various referenced objects and their relationships.

In this excerpt, Fruitloops has initiated a group practice of *indexicality* using the alphabetical labels of GeoGebra points to reference the objects formed by the points.

> *Group collaboration practice #11: Indexicality.*

Actually, this was such a natural practice for the students that Cornflakes initiates it simultaneously with her line 207: "ok so we have a poly gon with points k,e,s,t,l,i." In fact, Cornflakes starts to type her posting two seconds earlier than Fruitloops, although it is displayed two seconds later. Cornflakes is even more complex, referencing a polygon with the labels of its six points (including Fruitloops' two points). Note, however, that Cornflakes does not refer to "polygon KESTLI" the way that Fruitloops refers to "line KL," so that hers is more a describing than a pointing.

Cornflakes takes control and does steps 1 to 4, creating points and a segment (line 187). Then Fruitloops takes over and extends the segment into a quadrilateral. When Cornflakes now calls it a polygon, Fruitloops specifies that it is a "quadrilateral- a four sided shape?" Cheerios asks, "isnt it a rectangle?" but Cornflakes points out, "no its not syymmetrical." Cheerios is reacting to the current rough appearance of the shape as looking rectangular. In addition to ignoring the fact that it is not exactly rectangular because, for instance side KE does not appear to be equal to side LI, Cheerios is ignoring the ability to change the appearance by dragging the vertices.

Here we see a pattern of interaction in the group being repeated: Cornflakes does an interesting construction in GeoGebra. Fruitloops then refines the description that Cornflakes used from "polygon" to the more specific term, "quadrilateral." Cheerios tries to continue the refining discussion by suggesting it be called a "rectangle"—and presenting this in question format to elicit follow-up confirmation or correction. Cornflakes asks Fruitloops what a quadrilateral is, and then before getting an answer to that responds to Cheerios that the polygon cannot be a rectangle because it is not symmetrical, but that it is a quadrilateral. Fruitloops provides a definition for quadrilateral, but also hedges it with a question mark.

The team is hesitantly beginning to discuss the mathematics of figures that they construct. This involves the use of mathematical terms. As novices in geometry, the students are not knowledgeable or confident in their use of the technical terminology. However, by repeatedly using the words and discussing them in the group, they synthesize the understandings of the different students and begin to develop a richer sense of the application of the terms in various settings. People generally learn new words, including technical terms, in this way: starting to hear and to use them in contexts that give them gradually elaborated meaning (Sfard, 2008b; Vygotsky, 1934/1986; Wittgenstein, 1953), rather than by memorizing explicit definitions. This kind of mathematical discourse is a central goal of the VMT project. There are many prompts in the instructions intended to encourage the practice of *using new math terms* in the chat. Here, we see the team starting to enact this.

> *Group collaboration practice #12: Using new terminology.*

Fruitloops also creates two circles (line 200), in accordance with step 6 of the instructions. She drags the points on the circumferences to change the size of the circles. Next, she makes two points on segment KL (line 205). She drags one of these points to see that it stays on the segment even as it moves up and down it. (This introduces an important group practice in mathematics—

specifically dragging—but we will start to track such practices in the next chapter, where they become more the focus of group attention.) Recall that Fruitloops had similarly constructed a point on a segment in the first tab and dragged it along the segment. She also selected the compass tool, as requested in step 8, but she did not construct anything with it. During this session, no one investigates the dependencies created with the use of the compass tool. After her explorations, Fruitloops says, "someonee else explore" (line 218) and Cheerios finally volunteers to take control and do some extensive GeoGebra activity for her first time.

Cornflakes and Fruitloops had skipped over steps 5 and 7 involving dragging the example figures. Cheerios now undertakes this. First, she drags the circle with point G in response to step 7. However, rather than dragging point G around the circumference of the circle to visualize how the circle is made of points, she clicks on the center point and drags the circle as a whole around the screen. Similarly, she drags the circle CD around the screen.

Cheerios does not comment about her dragging at all. Eventually, Fruitloops asks her, "okay, walk us through what you're doing" (line 220). Over a minute later, Cornflakes repeats the request. Meanwhile, Cheerios constructs a new figure of her own design. She makes a segment and places a point on it. Then she constructs another segment starting at that point and forming an angle with the first segment (line 222). This leads to a discussion of complementary angles in the static view of the un-dragged figure. Meanwhile, in response to step 5, Cornflakes drags point J back and forth along the segment it is confined to.

After the team creates the three figures with connected segments or circles and points confined to them (see **Figure 7**), they decide to move to the next tab, Dragging (see **Log 9**). Fruitloops proposes that they go to the next tab. However, Cornflakes does not know if they should just do that or if they need to ask for permission, presumably from the teacher who is physically present in the classroom.

Log 9. The team decides to move on.

254	56:14.2	fruitloops	want to go to dragging
255	56:17.4	fruitloops	?
256	56:18.9	cornflakes	what?
257	56:26.8	cornflakes	whats dragagin
258	56:27.9	fruitloops	next taabb?
259	56:31.1	cheerios	it is a polygon
260	56:35.6	cornflakes	should we ask first
261	56:51.8	fruitloops	do vwe need to? i dont know

262	56:55.6	cornflakes	idk
263	56:59.9	cornflakes	not suuure
264	57:00.8	cheerios	im not sure
265	57:07.6	fruitloops	lets ju8st go
266	57:18.1	cheerios	okay so dragging it is
272	57:36.8	cheerios	dont forget to be descriptive
273	57:42.1	cornflakes	k

None of the three students is sure if they need to ask for permission. However, Fruitloops proposes that they just go without asking (line 265). Cheerios responds by declaring the decision made and Cornflakes agrees. Here we see the group deciding to take control of their own activity. The decision to move on is reflected upon by the whole group and then decided on by the group. This displays an increase in the *group agency* (Charles & Shumar, 2009; Damsa, 2014) of the team.

> *Group collaboration practice #13:*
> *Group agency.*

Previously, the team had made some decisions about what they should do as a group and how to do them. But the decisions either happened as a result of individual actions or proposals by individuals. Here, the decision to go to the next tab is made through discussion and agreement by the whole team. Group agency is important for effective collaboration. It allows the team to make its own decisions on its action trajectory: what to do, when and how. Then, all participants are aware of what the team is doing; they have a shared sense of its meaning and they can more easily maintain co-presence, joint attention and shared understanding.

We now follow the team's decision to switch to the next tab.

Tab Dragging

The team moves on to the tab called "Dragging" (**Figure 8**). The figure that includes points A, B, C, D, E, F, G (and that is still visible in this screenshot taken at the end of the team's work) was included in the tab originally, before the team entered, as part of the instructions. The students constructed the other figures. This example figure was included in the tab to illustrate points that are "free" (e.g., endpoints of an isolated line segment), "constrained" (e.g., points F and G confined to move along a segment or circle) or "dependent" (e.g., point E at the intersection of two segments). The instructions are designed to step the group through constructing a figure that includes these three different kinds of points.

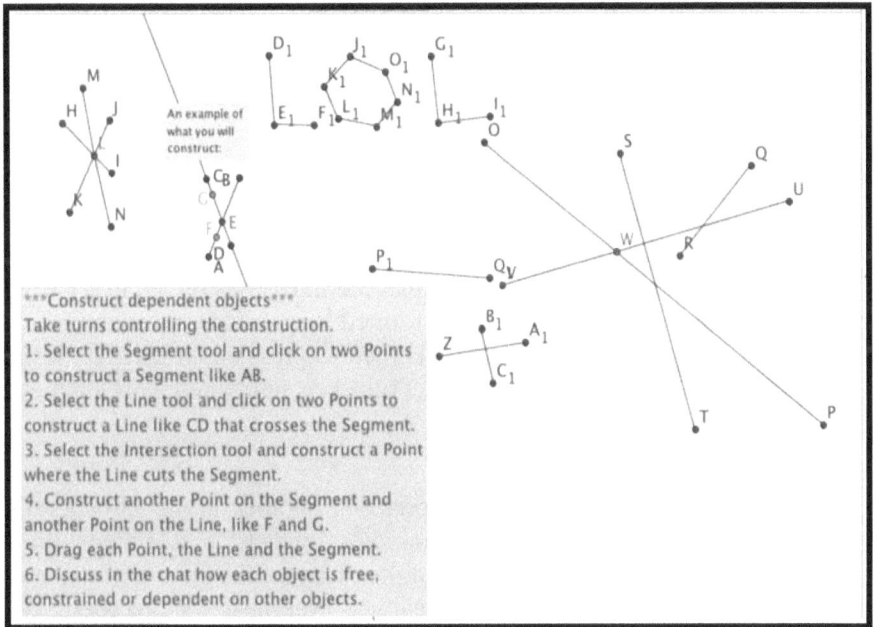

Figure 8. The Dragging Tab.

Log 10. Discussion of crossed lines.

274	58:01.2	fruitloops	im following the intructions
275	58:07.3	cheerios	okay
276	58:11.8	cornflakes	what step are you on?
277	58:25.0	cheerios	describe what you are doing
278	58:56.6	cornflakes	tell us the step your on
279	59:00.8	cheerios	yeah
284	59:56.7	cornflakes	fruity whatcha doing
285	00:05.2	fruitloops	i just tried to construct a line lioke the example buit i failed
286	00:12.4	cheerios	can i try

287	00:16.8	fruitloops	sure
288	01:20.4	fruitloops	how is each objettc free constrained or dependent?
289	01:22.6	cheerios	how do u delete lines
290	01:23.2	cornflakes	so we have to make a line then make another line that crosses the segment that we just made
291	01:41.4	fruitloops	try it corn
292	01:47.5	cheerios	do u guys see the 4 lines that i made
293	02:07.6	fruitloops	yes

The team succeeds in constructing a set of segments (connecting points H through N) that looks similar to the given example—as well as some other connected segments (see **Log 10**). Note that the lines constructed near the end of the session by Fruitloops—at the top formed by points D_1 through O_1—can be read as the letters "LOL," a well-known expression in chat or instant messaging: "lots of laughs." Once more, Fruitloops has ended work on the tab with playful construction.

The introduction of playfulness represents another group collaboration practice, part of the dimension of *sociality*, which is a foundation of collaboration (Barron, 2000). Throughout the sessions, the girls use humor and friendliness to reference each other ironically as "yes ma'm" or "my peer." They are playful in their exploration of the GeoGebra tools, such as constructing "LOL" here. They are respectful of each other, asking permission or inquiring about agreement. These are social practices that are sometimes prompted for by the instructions and that are ubiquitous in social discourse. They contribute to smooth collaboration.

> *Group collaboration practice #14: Sociality.*

Both Fruitloops and Cheerios follow instruction steps 1, 2 and 3—although they do not announce the steps in the chat. It is noteworthy that by now Cheerios is taking a major role in the constructions.

Fruitloops and Cheerios do not differentiate in geometry between finite "segments" and "lines" (that continue in both directions indefinitely off the screen), which are created by different tools in GeoGebra. Both students succeed in constructing segments that intersect and in constructing a point at the intersection. They do not seem able to place a point along the segment, as instructed in step 4. Note that step 4 does not specify what GeoGebra tool to use, as the previous steps do.

In **Figure 8**, we can see several constructions of intersecting line segments with dependent points at some of the intersections. Each of the students on the team made one of these. (The figure with points A through G was given with the instructions.) First Fruitloops created the figure with points H to N. Then Cheerios constructed the figure with points O to W. Finally, Cornflakes did the simple one with points Z to C_1. Later Fruitloops drew the LOL configuration. Everyone saw what each other had constructed (see lines 292 and 293).

Significantly, everyone's work in this tab remained displayed in the tab, along with the original example. It seems that each student arranged her work so it would fit in the visible tab's space without interfering with the work of the others. Then, people could have an overview of this sequence of work and draw over-arching conclusions, as in line 305 of **Log 11**.

Log 11. Discussion of dependent points.

305	06:17.5	cheerios	ok thats good all the lines inresect at least through another line
306	06:20.4	cornflakes	duh
307	06:35.9	cornflakes	yes which was the objective of step3
308	08:18.8	cornflakes	so constraints are like restrictions
...			
331	10:44.7	fruitloops	i dont think they are dependant on eachother'
332	11:02.0	cornflakes	thats creative use of math
333	11:03.1	cheerios	they arent dependent

In effect, the visual workspace of the tab serves as a group memory. It is a space in which work on the team's task is displayed persistently and maintained as accessible for review. It functions as a "joint problem space" (Teasley & Roschelle, 1993) for the team's efforts—a repository of concepts and artifacts of their work together. It is a visual embodiment of what Hanks (1992) calls "the indexical ground of deictic reference" or Clark and Brennan (1991) call "common ground."

At this point in their session, the team has adopted the use of the computer screen as a joint problem space. This is a group practice that supports the team's *intersubjectivity*, much as ones mental memory supports ones subjectivity. Such intersubjectivity is closely connected to co-presence and joint attention (Stahl, 2016a). The team is present together in a shared world, attending to the same objects, jointly experienced through group meaning making, shared understanding, group cognition, group agency and group practices. The practice of using the shared GeoGebra workspace and shared chat panel as a group memory and visual joint problem space to support intersubjectivity supports the team's co-presence during longer sequences of interaction.

> *Group collaboration practice #15:*
> *Intersubjectivity.*

The team then discusses the dependencies, as specified in step 6. However, they are silent on step 5, which says to drag each point, line and segment. It is this kind of dragging—which they skip—that would show them the difference in behavior of objects that are free, constrained or dependent on other objects.

As we see in **Log 11**, the team is, nevertheless, starting to discuss dependencies. They have just barely begun to discuss dependencies here, but will discuss them in increasingly greater depth in the future. In later sessions, we will track the team's learning about dependency as a central thread within their mathematical learning and group practices.

At first (line 288), Fruitloops simply repeats the wording of step 6 from the instructions in the Dragging Tab, and no one responds to her question immediately. Then—perhaps based on the experience of constructing intersecting lines, which constrain each other's movements—Cornflakes says, "so constraints are like restrictions" in line 308. Soon, Fruitloops states (line 331), "i dont think they are dependant on eachother" and Cheerios agrees (line 333) "they arent dependent." Unfortunately, it is not clear (to their teammates or to us) from what they say which objects they are discussing or the basis for their opinions. In fact, they have not dragged any of the points in this tab—either points in the given example figure or in their own figures. So they have not here observed constrained or dependent dynamic behaviors. Their group collaboration practices are still fragile and fail to adequately support coming to shared understanding of dynamic-geometrical dependencies in their figures during this first session.

Their hour is over and the team logs out.

Summary of Learning in Session 1

In this chapter, we have concentrated on the team's collaboration. By noting how the students seem to be at a loss about how to work together in the opening minutes of their first session and then seeing how relatively productive they become as a team as the session progresses, it is clear *that* they learn much about how to collaborate. The question then is: *how* did they learn?

We have noted during our review of the session a series of group practices that the team adopted, which improved their ability to collaborate effectively. These practices are foundational for supporting collaborative knowledge building through the co-construction of social order by the virtual math team. We have labeled a number of observed *group collaboration practices* as follows:

1. *Discursive turn taking*
2. *Coordinating activity*
3. *Constituting a collectivity*
4. *Sequentiality*
5. *Co-presence*
6. *Joint attention*
7. *Opening and closing topics*
8. *Interpersonal temporality*
9. *Shared understanding*
10. *Repair of understanding problems*
11. *Indexicality*
12. *Use of new terminology*
13. *Group agency*
14. *Sociality*
15. *Intersubjectivity*

At first glance, this list may seem like an arbitrary collection. However, these practices have been identified and discussed frequently in the study of collaboration. Many of the practices have been identified in the context of the evolution of the human species into an extraordinarily social form of beings, who can live in large communities, hunt with coordinated strategies, teach tool making, pass down cultural traditions and communicate in complex languages (esp. #2, 5, 6, 9, 15). Such practices have also been shown to be central to how young children develop within extended families. Conversation Analysis has identified several of these practices as central to human communication generally (esp. #1, 3, 4, 7, 10, 11). Finally, these practices have been recognized or speculated about in studies of collaborative learning, including in CSCL research (e.g., #6, 8, 12, 13, 14, 15). However, here they are here all displayed in the recorded interaction of a single group, as it first enters an exemplary CSCL setting. The list is not exhaustive or universal. Different groups— consisting of different people with different tasks and different technologies leading to their own, unique, situated interaction trajectories—would have generated somewhat different lists of group practices, in distinctive ways as they learn how to collaborate.

This is *how the Cereal Team learned to collaborate*. It learned by adopting this complex of practices, one at a time. While they may have existed in some form as individual practices or as community practices, they were here enacted as group practices, through group interactional mechanisms, such as proposal, negotiation and agreement.

The team was guided in various ways to adopt many of these practices—by the topic instructions, by their teacher's advice, by their own past experiences both online and face-to-face, and by general community standards. However, each of these prompted approaches or general social practices had to be enacted within the team situation in the online VMT environment. We have observed this enactment as the students made explicit, voiced and displayed the approaches to each other. We have highlighted a number of such enactment occurrences in the preceding review of the interaction log. The instances we have pointed out are just examples. The establishment of a group practice proceeds through multiple repetitions and variations. Groups acquire practices like individuals acquire habits. This takes place through iteration and adaptation, including both backsliding and evolving.

Such learning is not an accumulation of "ideas" as rationalist theories would have it or of propositions stored in the mind as information-processing theories conceptualize it, but is largely the enactment of practices. The students did not accumulate propositional content about how to collaborate, store it in something analogous to a computer memory, which they can access and recite explicitly on demand. What people offer when asked to state what they have learned has more the character of retrospective rationalizations based on folk theories, assumed frameworks or theoretical prejudices. They are narratives designed to respond to the situation within which one is questioned, rather than forms of expression of what is somehow stored "in the mind."

A virtual math team learns collaborative dynamic geometry by enacting various complexes of group practices. Once they have implicitly or explicitly agreed to adopt a procedure as a group practice—like agreeing to take turns on numbered action steps—they more or less follow that procedure, without having to negotiate it again each time. In this book, we are documenting the enactment of such practices by the Cereal Team.

The impetus for new group practices is stimulated from multiple sources. Many communication practices are sedimented in the natural language (a contemporary American middle-class teenager dialect of English) that the students bring with them to the chat environment. Other practices come from their school or larger cultural environment. Each of the students contributes unique perspectives from their personalities, which are often mimetically picked up by the others—as will be seen in later chapters. In addition, the VMT environment with its dynamic-geometry topics and their instructions is carefully designed to guide student and group development. As the VMT project team discovers through analyses like this book what practices contribute to mathematical group cognition, they add or refine scaffolding in the environment to encourage the development of those practices.

The idea of teaching people an idea by explicitly telling them the idea as a proposition is based on commonsensical folk theories of learning. In the VMT project, we provide an environment in which teams of students are supported in adopting helpful group practices, which will contribute to the team learning collaborative dynamic geometry. Among the supports are explicit statements of suggested procedures in topic instructions, but these must be enacted as group practices to be effective. It is the interplay of explicit and implicit—of propositional instructions and hands-on exploration—that is effective, but hard for designers of learning environments to predict. The analysis—through detailed observation of team displays—of the development of the team's mathematical cognition—as the enactment of specific group practices—serves to guide iterative re-design of the learning environment.

At the end of the first session of the Cereal Team, we see some ways in which the group has begun to form itself into an effectively collaborative team. The students have adopted group practices that will remain with them. Probably the major advance during Session 1 has been in the area of collaboration, although the team has also had a first experience in using dynamic-geometry tools. Learning has taken place in the intended areas:

i.   The Cereal Team has adopted some basic *collaboration practices*, such as addressing each other in text-chat discourse, listening (reading) and responding. The students take turns, not only in the chat, but even more in GeoGebra actions. They first discuss who should take control, then sometimes describe what they have done and finally release control for someone else. Perhaps most significantly, they decide to follow the numbered steps of instruction in the tabs. When they do not know what someone is doing or understand why they are doing it, they ask a question in the chat. They use the chat to negotiate decisions for the group and to register agreement or disagreement. In general, they maintain a friendly atmosphere and are often playful, for instance in addressing each other with mock formality. They keep track of the time they have left to work and try to move through the several tabs for the session. None of this is perfect or fluid yet. However, their initial sense of not knowing what to do is quickly diminished and they are able to make progress through the topic.

We can see to some extent *how* the team develops as a collaborative group. They start out as individuals reacting to the online situation in which they find themselves. As they begin to act, they run into difficulties or breakdowns in the smooth functioning of their activity. They selectively take into account guidance offered by the topic instructions. They engage in group discourse and interaction, in which they elicit proposals for solutions to their quandaries. Gradually, offered proposals lead to the adoption of group practices, which all group members accept and which tend to overcome their difficulties. These collaborative practices include discussing who should do GeoGebra actions next, following numbered steps in the instructions and discussing what they have done. These practices were sometimes suggested in the instructions, but had to be enacted by the group—that is, adopted in specific ways within the team's concrete situation.

We have identified a number of *group collaboration practices* adopted by the team, which are general collaboration practices in society. They include many of the fundamental preconditions for productive interaction and work together. In particular, these group collaboration practices form the foundation for computer-supported collaborative learning and are central to a theory of CSCL and more generally of group cognition.

ii.   The team has taken initial steps in developing *productive mathematical discourse*. They have discussed math terminology and instructed each other on the meaning of several geometry terms, like "quadrilateral" and "complementary." They have begun to discuss the notions of constraint and dependency in a very preliminary way. Just as Cornflakes adopted the term "quadrilateral" from Fruitloops without being able to define it, so the team uses the ideas of constraint and dependency tentatively, without confidence in understanding what they mean precisely.

iii.   The team has *learned to use GeoGebra's tools* for dragging and constructing simple dynamic-geometry figures, including connected segments and points confined to a segment. Each of the students has engaged in constructing and dragging GeoGebra objects consisting of points, segments and circles. Even Cheerios finally starts to drag objects, although she does not do so in a way that displays their dynamic character or their invariants and dependencies; she does not show any understanding of the dynamic character of the figures she is manipulating.

More specifically, we can track the team's initial fluency with *identifying and constructing dynamic-geometric dependencies*:

a. Some of the students have tried *dynamic dragging* of points. This is still quite tentative. They do not seem to have a strong sense of seeing the figures as dynamic; dragging is used more to position figures, which are still often observed in terms of their static visual shapes.

b. Each of the students has begun to engage in *dynamic construction* of simple figures, generally consisting of a couple of segments joined together. However, when they notice things—even at the end of the session—it is visual appearances of their static constructions, not the dynamic behaviors that the topics were intended to get them to focus on, like a point being confined to a segment.

c. The team has not begun to design *dynamic dependency* into GeoGebra constructions. They have not even commented on the simple dependencies observed during the few times that they dragged figures. Their discourse about dependencies is not yet along the lines intended by the design of the tasks. The team's understanding of dependencies is vague and still not informed by experience dragging dynamic-geometry objects, like points confined to a segment or to an intersection, or points shared by connected segments.

We can already begin to draw some preliminary *lessons for re-design* of the topics based on the observed behavior of the team in Session 1:

*   The first tab, Welcome, should be structured more clearly into a sequence of numbered tasks, where each task can be done by one person, and then tried by each of the other team members.

*   The second tab, Help Hints, should be made available to students before they enter into a team, to read and explore on their own. The zooming and other actions are not reproduced on everyone's screen, so it is impossible to follow what others are doing. The Welcome Tab should also be made available in advance as a warm-up or introduction for individual students to try on their own. This will give them more time to explore and play with the most basic tools. The tab can be used again in the first collaborative session so they can share what they have learned and get help from teammates for things they had trouble with. (This had actually been the plan in WinterFest 2013, but the teacher did not organize the warm-up individual sessions. More effort should be made to do this, although it can be difficult to motivate and organize.)

*   There should be more prompts or tasks encouraging students to announce in the chat what they plan to do in GeoGebra and then what they have done. If there is a possibility that students are sitting physically near to each other in the same room, they should be

encouraged to communicate only through the chat, so that there is a record of their collaboration.

- Although specifying numbered tasks to step through can be helpful in the beginning, generally there should be more explicit focus on the principles of dynamic geometry that are being explored than on the completion of specified tasks. The numbering should correspond to meaningful whole actions.

- The examples of dependencies—such as points constrained to a segment or to an intersection—should be highlighted and discussion of the dependency relationships should be more explicitly prompted. In general, the activities should be focused even more specifically and narrowly on the notion of dependencies.

- The difference between visual appearances of static figures and relationships of dependency in dynamic figures should be pointed out. Prompts for noticings should emphasize noticing things that remain true (invariant) under dynamic dragging.

While we have discovered that the VMT environment and pedagogical approach can be improved in a variety of details, it is also clear that the general strategy was effective. The team substantially increased their ability to collaborate effectively—in their first hour together online. This was not achieved by subjecting the team members to a verbal or written lecture on how to collaborate. Rather, the students were situated in a collaboration environment and were guided to work together in ways that allowed them to enact a variety of group practices, which laid a foundation for collaborative interaction as a group.

Session 2: The Team Develops Dragging Practices

The second session starts much like the first. The team expresses considerable uncertainty about how to proceed. However, they persevere, with each team member taking turns trying and encouraging the others. They learned in the first session to focus on the numbered steps in the instructions and they now proceed with that group practice. They try to follow the steps in the first tab, to construct an equilateral triangle (**Figure 9**).

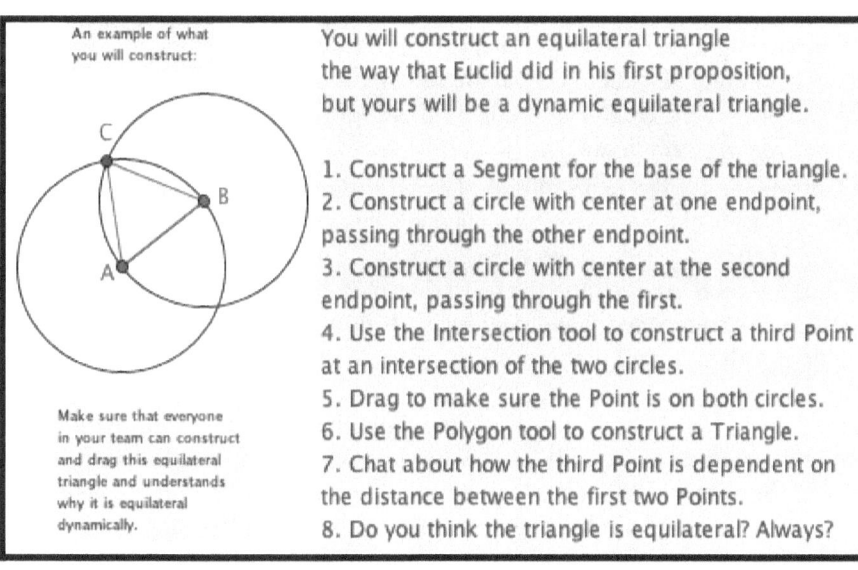

An example of what you will construct:

C
B
A

Make sure that everyone in your team can construct and drag this equilateral triangle and understands why it is equilateral dynamically.

You will construct an equilateral triangle the way that Euclid did in his first proposition, but yours will be a dynamic equilateral triangle.

1. Construct a Segment for the base of the triangle.
2. Construct a circle with center at one endpoint, passing through the other endpoint.
3. Construct a circle with center at the second endpoint, passing through the first.
4. Use the Intersection tool to construct a third Point at an intersection of the two circles.
5. Drag to make sure the Point is on both circles.
6. Use the Polygon tool to construct a Triangle.
7. Chat about how the third Point is dependent on the distance between the first two Points.
8. Do you think the triangle is equilateral? Always?

Figure 9. The Equilateral Tab.

We will now switch from highlighting group collaboration practices to focusing on group mathematical practices, beginning in this chapter with practices for dragging in dynamic geometry. Learning to work in GeoGebra requires considerable trial and practice. The students develop numerous practices of dragging and construction—too many to investigate in this book. The three students are able to pick up many of these practices on their own, as they manipulate GeoGebra objects or watch their teammates' efforts. In this chapter, we will concentrate on practices that are adopted by the Cereal Team as a whole in ways that are visible in their interaction. These may be practices that the individual students were not able to pick up on their own and had to be discussed and shared in the team. We will enumerate a number of these group practices as typical of how student teams can learn to engage in dynamic mathematics and develop their mathematical group cognition.

Tab Equilateral

As the instructions mention, this construction was the starting point for Euclid's presentation of geometry (Euclid, 300 BCE). It is a paradigmatic construction; a good understanding of it provides deep insight into the nature of Euclidean geometry. Similarly, the construction of an equilateral dynamic triangle in GeoGebra can convey the core of dynamic geometry: constructing, dragging, creating dependencies, establishing equalities of lengths, marking intersections and organizing a set of relationships to achieve dynamic behaviors. One cannot expect beginning students to grasp the full significance of this construction. We will see how the Cereal Team enacts this topic—by highlighting a series of *group dragging practices*.

Unfortunately, the team already has considerable trouble with the second step: "2. Construct a circle with center at one endpoint [of the segment constructed in step 1], passing through the other endpoint." (See **Log 12**.) The wording is perhaps a bit too cryptic, and the team does not try to make detailed sense of it (line 42). Although they have decided to follow the steps of the instructions, they do not always read them carefully or try to interpret their precise meaning. Reading closely and taking into account the precision of wording in mathematical text is itself a mathematical practice that the group will have to gradually acquire (as we will see). Now, instead, they proceed to create many objects, seemingly without much planning. They spend a half hour constructing points, segments and circles before managing to accomplish step 2.

Log 12. The team tries to construct an equilateral triangle.

40	22:19.1	cornflakes	fruitloops use de
41	22:29.0	cheerios	wheres the circle
42	22:37.5	fruitloops	okay but i dont understand step 2
43	22:39.6	cornflakes	make a triangle and attach 2 circles
44	23:17.7	fruitloops	like d, f, e?
45	23:21.7	cornflakes	yes
46	23:30.2	cornflakes	fruitloops make the circles bigger
47	24:15.4	fruitloops	someone else take control

| 48 | 24:19.3 | cornflakes | delete the triangle! |
| 49 | 24:52.5 | fruitloops | done! now someone else |

Cornflakes starts by making a line segment DE (line 40) in response to step 1. But then no one knows how to proceed. They know to create circles, but they do not seem to understand that the endpoints mentioned in step 2 are the points D and E at the ends of their new segment, DE. Cheerios constructs a series of circles and drags their centers and circumference points to explore them and relocate or resize the circles. Cornflakes also makes a number of circles, without attaching them to the segment. The team seems to need to explore the nature of constructing circles and associating them with lines or points before it can succeed in following the instruction steps. Learning to work in dynamic geometry requires considerable playing around.

It is clear to the team at this point that progress in working on the given task of constructing an equilateral triangle will involve learning how to use a number of GeoGebra tools. The instructions of the task refer repeatedly to constructing and dragging, as well as specifying use of the intersection tool for one of the steps. From the previous session, the students know that construction and dragging in this environment involve the selection and use of specific tools from the GeoGebra tool bar. In particular, steps 2 and 3 explicitly call for the construction of two circles—as illustrated in the geometric figure that is already shown in the tab. The task requires a rather precise usage of the circle tool; the instructions try to describe this usage in some detail. The use of GeoGebra tools in the previous task was less specific—students could create points and segments more freely and be satisfied with whatever resulted from their inexperienced usage of the tools. In this topic, the students must master the usage of the circle tool, which is more complicated than the point or segment tool, as they discover.

At one point, the students start with a triangle and then try to add circles to it to make it look like the example figure (line 43). At other times, they create circles and try to adjust them to look like the example figure. This suggests that the students are basing their work on the static visual appearance of the figure, rather than on the dynamic relationships of equal radii.

Figure 10 shows an example of their work shortly before finally succeeding. We can see a couple of the many points, segments and circles the team has created in trying to reconstruct the given figure of an equilateral triangle based on two circles with radii of the base side of the triangle (segment DE). Cheerios connected the two circles to endpoint E on segment DE, but not to point D—so the team is starting to approach the solution: They have "a circle with center at one endpoint" and another circle "passing through" that same endpoint. In particular, at the moment shown, one endpoint of the base segment, DE, is serving as a center of one circle and a point on another circle. So, the students are finally getting close to the called-for construction, but are still rather confused.

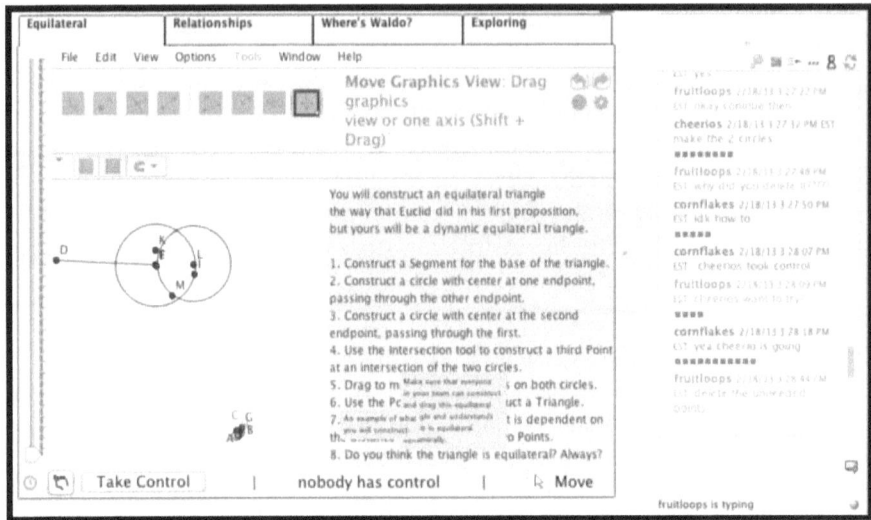

Figure 10. An attempt to construct an equilateral triangle.

Cornflakes builds on the overlapping circles from Cheerios. She adds a point J on one of the circles near (but not even visually at) the intersection of the circles. She then draws a triangle connecting points E, L and J. Then she hands control over to Fruitloops. However, Cheerios takes control instead and erases the triangle, commencing to try more points and circles.

Fruitloops does not know how to proceed. While Cheerios is creating and erasing points, Fruitloops asks Cornflakes to take control. Cornflakes clears the workspace and constructs once more a base segment AB. Fruitloops likes that (line 82), and she guides Cornflakes through the chat (**Log 13**). Perhaps when she saw the workspace cleared off of all the false starts and now containing just a segment AB, Fruitloops could see how to proceed. She reformulates step 2 and 3 as "now make a circle from both endpoints" (line 83). Cornflakes constructs a circle centered on B and with circumference defined by A. Now she seems to understand the involvement of the "endpoints"; students generally understand the instructions of their peers better than textual instructions.

Log 13. Constructing the circles.

| 80 | 32:33.4 | fruitloops | what should i do? |

81	32:43.5	fruitloops	coernflakes take control
82	33:43.2	fruitloops	yes thats good so far
83	33:58.2	fruitloops	now make a circle from both endpoints
84	34:11.8	cornflakes	cheerios take control
85	34:17.3	fruitloops	right?
86	34:48.4	fruitloops	cheerios go
87	34:56.3	fruitloops	do you understand what to do?
88	35:00.2	cheerios	im not sure how to do it
89	35:21.7	fruitloops	dont you have to make a circle from point b?
90	36:58.9	fruitloops	why did you makee your point c there?
91	37:28.5	fruitloops	okay nevermind
92	37:32.9	fruitloops	make point c now

For a minute, no one knows how to construct the second circle. Cornflakes and Fruitloops invite Cheerios to try, but she is also not sure how to do it (line 88). Finally, Fruitloops suggests, "dont you have to make a circle from point b?" (line 89). So Cornflakes selects the circle tool and clicks on point B as the center. However, instead of clicking on point A to define the circle going through it, she clicks on a location about half way between B and its circle, creating a new point C and a circle around B through C. Before releasing point C, however, Cornflakes drags it until the circle that it defines visually looks like it is also passing through point A.

Fifteen seconds later, Cornflakes deletes the new point C along with the new circle. She then does the same thing with a new point C to the right of B. Again, during its creation the circle through C is dragged to appear to go precisely through point A. Although it looks like the two circles are both defined by the endpoints A and B, the new circle is defined by A and C. The radius of the circle is not defined to be dependent upon segment AB. It merely looks like it passes through B now, but if any points are dragged the circle will no longer pass through B. Cornflakes does not do a drag test to check this. She has adopted a practice that produces a circle that looks like it involves the target points A and B, but unfortunately will not withstand the drag test and therefore is not a valid practice in dynamic geometry.

This invalid practice is like a "student misconception" in that it may be necessary for the students to pass through the stage of trying this practice and discovering that it does not hold up in the dynamic-geometry simulation—in order to advance to trying a somewhat different practice that will prove to be valid. In this sense, it is a temporary and flawed, but important, step by the group in learning about construction and dragging in dynamic geometry.

We can see in all this trial-and-error work that the students have yet to grasp a fundamental principle of dynamic-geometry construction. Constructions must be built in ways that define relationships among the involved objects (points, segments, circles, etc.). The equilateral-triangle construction, for instance requires that one circle be defined as centered on point A and passing through (i.e., created with) point B. Point B has to be used in the construction; the circle has to be defined in terms of B, not just happen to pass through it. Only that way can the software maintain the condition that the circle passes through B. Otherwise, when one drags A, B or the circle, the circle will move away from B. In watching the students, we can see that this principle is by no means obvious and takes a major insight based on exploration. Grasping this principle by changing how they construct circles will be an important step in learning to do dynamic geometry. The students engage in some discussion of how they are defining their circles.

Fruitloops asks, "why did you makee your point c there?" (line 90). This may imply that there is no reason why Cornflakes should create a new point instead of defining the circles using points A and B. Meanwhile, before Fruitloops' message is posted, Cornflakes again deletes point C along with its circle. She then constructs the circle around B and through A. Both Fruitloops and Cornflakes see that the circle has to be constructed using the point B to define its circumference, rather than using an arbitrary new point and then adjusting its position to make the circle seem to pass through B. Fruitloops gently suggests this with her inquiry about making point C, but Cornflakes has apparently also realized it on her own. All the trials that the group has gone through have made this clear.

The students already have a variety of individual practices they have tried for constructing and dragging geometric objects in the GeoGebra environment. However, now they have jointly adopted a first important *group dragging practice* which captures the spirit of dynamic geometry:

> *Group dragging practice #1: Do not drag lines to visually coincide with existing points, but use the points to construct lines between or through them.*

Fruitloops says "okay" to the new construction and then suggests Cornflakes "make point c now" (line 92), meaning the point of intersection of the two circles. Cornflakes turns control over to Fruitloops (line 93 in **Log 14**), who actually constructs the triangle by locating point C at the intersection of the two circles (step 4), with direction from the others (**Figure 11**). Although Fruitloops seems to understand how to construct the circles and their intersection, she directs her teammates to do the actual manipulation in GeoGebra. When it is her turn, she does not seem to know how to construct a line connecting two points by using the GeoGebra segment tool: "how do i make the points connect?" (line 94). This is reminiscent of her question in line 80 of **Log 5** in Session 1, which she was immediately able to resolve by herself. When Cornflakes and Cheerios tell Fruitloops to use the line segment, she constructs point C at the intersection of the circles so that she will be able to connect the vertices of the equilateral triangle.

Log 14. Connecting the points.

93	37:42.6	cornflakes	take control
94	38:07.2	fruitloops	how do i make the points connect?
96	38:49.8	cornflakes	yao line segment it
97	38:56.5	cheerios	line segment

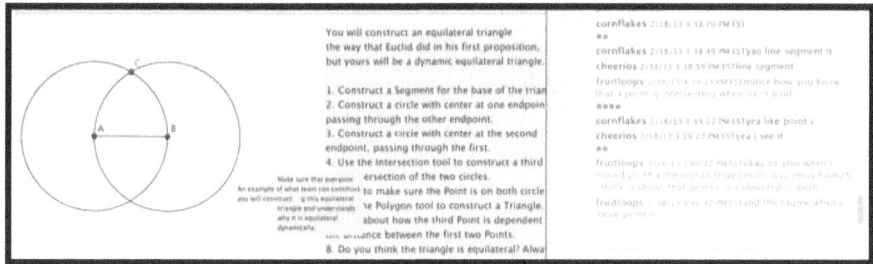

Figure 11. Intersecting the circles at point C.

It may be that Fruitloops was not asking about what tool to use, but planning aloud the need to actually define a GeoGebra point at the intersection of the circles in order to connect up the vertices of the triangle, which include the intersection. According to Vygotsky, self-talk is an intermediate between discourse and silent thought (Vygotsky, 1930/1978; 1934/1986). It is interesting to consider Fruitloops' posting as something analogous happening in online chat: a query directed primarily to herself, but also displayed to the others. Sharing such self-talk is an important step in shifting from private thinking to conducting problem solving collaboratively.

It seems that none of the students is able to do the tasks on their own; they each have a partial and growing understanding, which gets articulated enough to guide the accomplishment of the task through their interaction. Constructing the equilateral triangle in GeoGebra is not a straight-forward matter of just reading some instructions and using the tools to do it. One must learn how to read geometry-construction instructions and how to use GeoGebra tools to create geometric objects that accord with the instructions. For instance, to "construct a circle with center at one endpoint passing through the other endpoint" is tricky. One must select the circle tool and click on one of the endpoints of the segment to define the circle's center first. Then one must click on the other endpoint to specify that the circumference of the circle goes through that point. One cannot first create a circle somewhere and then drag it to go through the points. This is practical knowledge that one must gain through practice with GeoGebra. The students gain such knowledge as a group by watching how each other eventually succeeds and by guiding each other to follow the effective construction practices.

Fruitloops points out that the GeoGebra system indicates that she constructed point C at the intersection of the two circles by making the circles both appear thicker or in bold to show that they were selected by the cursor placing the point (line **98** in **Log 15**). This is an indication by the GeoGebra software that point C is being defined in terms of the two highlighted circles. Cornflakes and Cheerios both acknowledge that lesson. Here, the students are seeing that they have taken a construction action that is recognized by the software system. They are learning that using dynamic geometry involves interacting in specific ways with the software so that the system can support the construction's relationships (e.g., that a point is indeed being constructed at an intersection). The system often provides confirmatory feedback, such as making a point or a line temporarily bold.

> *Group dragging practice #2: Observe visible feedback from the software to guide dragging and construction.*

Log 15. Dragging the points.

98	39:13.8	fruitloops	notice how you know that a point is intersecting when its in bold
99	39:22.1	cornflakes	yea like point c
100	39:27.1	cheerios	yea i see it
101	40:27.7	fruitloops	okay so also when i moved point a the rest of thwe circles also moved which i think it shows that point c is connected to both
102	40:42.3	fruitloops	and the saqme when i move point b
103	40:46.7	cornflakes	okay yes

104	40:51.0	cheerios	yea so it is intersected both circles
105	40:52.2	cornflakes	i see thst
106	41:11.2	cheerios	now we have to make a triangle

Such system feedback is helpful—particularly in knowing when one has located ones cursor at precisely the intended location. However, it does not guarantee that all the relationships are established the way one wants them. For this, one must see what happens when one drags various objects. Are the intended relationships retained? Do the triangle's vertices remain at the segment endpoints or at the circle intersection? Checking that relationships in a construction are maintained dynamically by dragging objects is called the "drag test." Establishing the habit of checking constructions with the drag test is a fundamentally important practice. Fruitloops here checks her connection of point C to the intersection of the two circles.

Fruitloops does a quick drag test, following Step 5 of the instructions in the tab ("5. Drag to make sure the Point is on both circles."). This is the first time the students are using dragging to determine dependencies among constructed objects. First, she drags point A a small distance (line 101) and then point B a short distance (line 102). In both cases, the circles move in a way that maintains all the relationships, including that point C stays at their intersection. Again, Cornflakes and Cheerios both agree with this important observation. This reflects recognition of the need to make the construction valid in a dynamic-geometry sense. The team thereby moves beyond its earlier misconception.

The students have all seen and acknowledged the importance of the drag test, which was prompted by the instructions, but which came alive in the context of their work together. In conducting the drag test for herself, Fruitloops has displayed the validity of the construction to the whole team. The others display their alignment with her display through their chat postings (lines 103, 104 and 105). This establishes their group dragging practice involving the drag test:

> *Group dragging practice #3: Drag points to test if geometric relationships are maintained.*

They conclude that the construction is successful and that they can use the Polygon tool to draw in the desired triangle connecting these points A, B and C, as instructed in Step 6 (line 106). The team then moves on to Step 7 (line 109 in **Log 16**), which raises the issue of dependencies: "7. Chat about how the third Point is dependent on the distance between the first two Points." They discuss the question in various ways. Cornflakes responds in terms of the construction. It is not clear what distance she is referring to. If it is the distance from A to C and from B to C, then she is close to the main insight of Euclid's proof of equal triangle sides based on equal radii of congruent circles. Fruitloops makes the interesting observation that the triangle is *always* an equilateral, probably referring to its maintenance of relationships under dragging. Cheerios stresses that it is equilateral by definition of having equal sides—but that could be based on a non-dynamic view of the static shape. The team begins to use dragging to identify or test invariances in figures:

> *Group dragging practice #4: Drag geometric objects to observe invariances.*

Log 16. Dragging the triangle.

109	41:54.1	fruitloops	do you have any idea of how to answer 7?
110	42:29.7	cornflakes	the 3rd points dependent on the 1st 2 points because the kind of triangle it forms is dependent on thedistance
111	42:41.3	fruitloops	i think the traingle is always an equilateral. do you agree?
112	42:54.6	cheerios	yes it is because all sides are equal
113	42:58.4	cornflakes	yes cause the intersection
114	43:02.1	cornflakes	yea
115	43:14.8	cornflakes	lets mov eon
116	43:17.0	cornflakes	?
117	43:17.5	cheerios	correct because its right in the middle
118	43:27.1	fruitloops	yeah even when you move any of the points likie for example if i moved point b, the triangle stays equal.
119	43:36.9	cheerios	it always will be equaladeral
120	43:45.2	fruitloops	okay i agree.
121	43:49.0	cheerios	the triangle^
122	43:53.8	cheerios	lets move on
123	43:55.1	fruitloops	do you want to move on to relationships?

Cornflakes adds, "yes cause the intersection" (line 113) as support for Fruitloops' conjecture that the triangle is always equilateral. Thus, the team is aware of the equality of the side lengths and the fact that the construction of the third vertex at the intersection of the two circles is involved in making them equal. However, they never articulate the role of the circle radii. It is interesting that the students respond to the prompts about the point being "dependent" and the issue of relationships "always" being true with claims using the logical connective "because," although they are not able to put together a proof-like sequence of argumentation, including explaining how they know the sides are necessarily equal in length. By using causal terminology and providing evidence, they display a commonsensical or everyday-language form of argumentation, which can gradually be refined into a mathematical form of deduction.

The group uses its response to step 7 to help answer the final point of the tab: "8. Do you think the triangle is equilateral? Always?" The determination of the equality of the side lengths implies that the triangle is equilateral, by definition of "equilateral." Fruitloops answers the "Always?" part by referring to what she found when she dragged point B: "yeah even when you move any of the points likie for example if i moved point b, the triangle stays equal." (line 118). This reflects the important recognition of the significance of the drag test. Prompted by the instruction in step 5, the team has begun to use the drag test to verify dynamic construction dependencies, and here they articulate in their discourse this use of it.

> *Group dragging practice #5: Drag geometric objects to vary the figures and see if relationships are always maintained.*

The *drag test* is an aspect of dynamic geometry that many researchers and teachers of dynamic geometry view as fundamental to this form of mathematics (Arzarello et al., 2002; Goldenberg & Cuoco, 1998; Hölzl, 1996; Jones, 1997; King & Schattschneider, 1997; Laborde, 2004; Scher, 2002). There are several aspects to the role of the drag test. One is that it is a way to test whether a construction attempt is successful. For instance, by dragging points A and B, Fruitloops tested that point C remained at the intersection of the circles around A and B, confirming that her attempt to mark the intersection with point C worked (line 101). Another is to vary a geometric figure while maintaining the dependencies that were constructed into it. As the students drag the vertices of their triangle, its size, location and orientation change. In this way, the figure that they originally created in one position can visually take on and represent a large number of variations. The students can then see that certain relationships—like the equality of the side lengths—remain across all the variations (line 118). That leads them to say that the triangle is "always" equilateral. The ability to consider variations like this—promoted by experience dragging figures—may be considered an advance in van Hiele levels on the way to thinking in terms of proofs.

After their reflection on the construction of the equilateral triangle, the team moves on to the next tab.

Tab Relationships

The Relationships Tab builds on the equilateral triangle construction to add more related segments and angles (see **Log 17** and **Figure 12**). While Cheerios and Cornflakes start to describe the visual appearance of the complex figure (lines 128, 129, 130), Fruitloops notes its resemblance to the equilateral triangle construction (lines 132 and 136). Notice that all three students are oriented toward the intersection of the two large circles, which were important in the previous tab.

Log 17. Identifying constraints.

128	44:52.8	cheerios	well they are 2 circles that are intersectiong each other
129	45:29.5	cornflakes	two circels intersecting each other
130	45:39.5	cheerios	the space creates an oval
131	45:49.1	cornflakes	points e d and c are contrsined
132	45:53.2	fruitloops	it reminds me of the shape from the equilateral tab
133	45:59.5	cornflakes	right
134	46:04.0	cornflakes	theyv are similar
135	46:10.0	cheerios	yea it is because point d and e is black
136	46:12.4	fruitloops	except more points are added adding more triangles inside
137	46:25.4	cheerios	both of the triangle are equaladeral
138	46:34.4	cornflakes	point e is in the dead center
139	46:52.7	cheerios	yea its more complex because of the added line segments which make different polygons
140	47:22.2	cheerios	there are 4 isocles triangle which look like a large diamond
141	47:31.8	fruitloops	which points are free and which are constrained?

142	47:44.5	cheerios	each triangle make 2 acute angles and one right angle
143	47:47.5	cornflakes	e d and c are constrained
144	48:07.8	cornflakes	because you csant move them around they are conmnnected to multiple thangs
145	48:14.6	cornflakes	*things
146	48:33.5	cornflakes	does iy make sence
147	48:38.1	cornflakes	*sense
148	48:40.4	fruitloops	what about f?
149	48:47.6	cheerios	yes because it makes a shape
150	48:49.2	cornflakes	yeah f too
151	49:02.9	cheerios	and is connected to the shared vertices
152	49:16.9	cornflakes	yup
153	49:40.5	fruitloops	what about point e?
154	49:47.0	cornflakes	point e is smack in the middle
155	50:19.4	cornflakes	the colors of the poinjts indicte what they are'
157	50:24.3	cornflakes	contrained or whastever
158	50:24.4	fruitloops	and it doesnt move
159	50:38.6	cornflakes	yes
160	51:00.0	fruitloops	what segments are the same lenght?
161	52:07.7	cheerios	segments de and ec are the smae length
162	52:08.7	fruitloops	db and da and ba and bc and ac i think are the same lenghts
163	52:30.1	cheerios	yea they all are the same
164	52:48.0	cheerios	be and ea are the same
167	54:12.3	cheerios	they are all right angles
168	54:24.1	cheerios	90 degree angles
169	54:24.7	fruitloops	all the angless near point e are right angles
170	54:30.8	cornflakes	the angles near point e are right angles bcuz point e is located in the center
171	54:39.0	cheerios	what are conjectures
172	54:46.5	fruitloops	i dont know
173	55:28.8	cheerios	what are relationships are u guys unsure of
174	55:32.4	cornflakes	2 angles forming alinear pairr
175	55:38.6	cornflakes	i think

176	55:47.3	cheerios	which ones
177	56:07.0	fruitloops	Where's Waldo?
178	56:40.1	cornflakes	we have to answer 9 and 10
179	56:51.6	cheerios	yeah
180	57:02.6	fruitloops	i dont really know about 9.
181	57:09.3	cornflakes	meneither
182	57:32.5	cheerios	the black dot means that they are in the middle so the lines on either sides of it have to be the same length

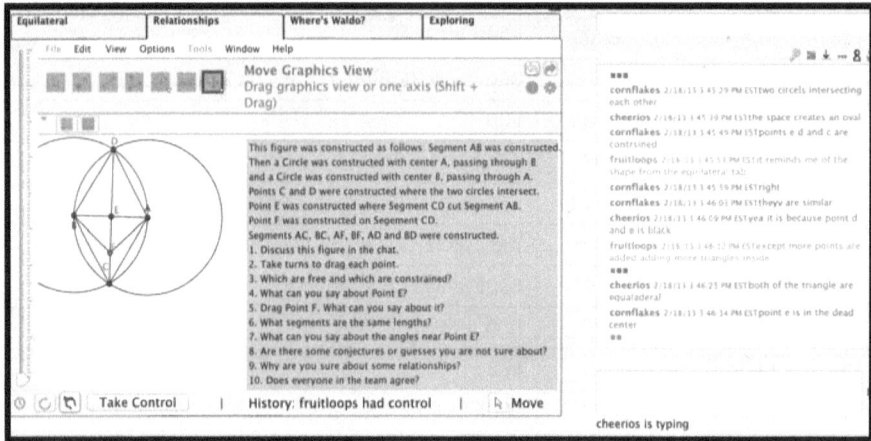

Figure 12. The Relationships Tab.

From lines 130 to 136 Cornflakes has control of the GeoGebra tab and drags the figure energetically. She especially drags points A and B, discovering that points C, D and E cannot be dragged directly: "points e d and c are contrsined" (line 131). Cheerios adds that "yea it is because point d and e is black" (line 135). This is a key observation, although the causality is confused. Points are colored black in GeoGebra to indicate that they are dependent on other objects, not to make them dependent. Although the color-coding of dependent points can provide a helpful clue to understanding GeoGebra figures, the reliance on this coloring often distracts students from understanding dependency relationships in terms of their construction. Cheerios mentions points D and E, but if she used the lesson from the previous tab, she might realize that points C and D are dependent because they are intersection points of the two circles. Nevertheless, the students have associated the term "constrained" with two consequences of being constrained: the ability to drag a point is limited and the point appears in a different color in GeoGebra. They are developing their discourse about dependencies.

> *Group dragging practice #6: Some points cannot be dragged or only dragged to a limited extent; they are constrained.*

Cheerios combines the visual description of the diamond-like appearance of a figure with identification of its geometric properties, like being isosceles in line 140: "there are 4 isocles triangle which look like a large diamond." Cornflakes explains the constraints on the points as being "because you csant move them around they are conmnnected to multiple thangs" (line 144). Cheerios affirms this: "yes because it makes a shape" and "and is connected to the shared vertices" (lines 149 and 151). This reflects that it is the geometric connections among points and lines in the figure that accounts for the dependencies. However, there is no detailed accounting of why certain lengths are equal to each other, resulting in the "shapes" being isosceles or equilateral. While Cornflakes tries to explain things in terms of their construction and geometric relationships, Cheerios repeatedly reduces the discussion to visual, static shapes.

Fruitloops explores the dynamic relationships or constraints in the construction through dragging. From lines 138 to 144, Fruitloops has control of the GeoGebra tab and drags the figure energetically. She especially drags point F in response to step 5 of the instructions. In line 148, she asks "what about f?" Cornflakes responds that point F is also constrained (line 150). Fruitloops has just been dragging point F along segment CE and this has been visible to the whole team. No one remarks that F is constrained to move along a segment, whereas C, D and E can not be dragged at all, due to their definition as points of intersection. The team has not noted this distinction between being partially constrained and being fully dependent upon other objects.

Many of the team's explanations are descriptive of visual appearances. When confronted by step 9's question, "Why are you sure about some relationships?" the team does not know how to respond, and decides to move on to the next tab. Their discourse shows no characteristics of proof-type arguments for the necessity of relationships.

Tab Where's Waldo

The Where's Waldo Tab reproduces the figure from the Relationships Tab, simply shading in the equilateral triangle ABC (**Figure 13**). It asks the students to identify different kinds of triangles within the larger figure. The team names various triangles and even corrects the tab's use of the term "scalar" in place of "scalene" (line 208 in **Log 18**).

Log 18. Identifying kinds of triangles.

197	59:50.7	fruitloops	there is definitly a right angle
198	00:01.1	fruitloops	right triangle*
199	00:10.5	cornflakes	yes there is 2
200	00:16.5	cheerios	how do u make it bigger
201	00:20.7	cornflakes	aef and ebf
202	00:30.3	fruitloops	4 right triangles right?
203	00:54.7	cheerios	yea
204	00:55.2	cornflakes	yes
205	01:12.5	cheerios	there are also isoceles and scalene triangles
206	01:31.6	fruitloops	what about scalar?
207	01:40.0	cornflakes	scalár?
208	01:42.5	cheerios	that is scalene
209	01:49.1	fruitloops	yeah
210	02:21.1	cornflakes	ya
211	02:40.8	cheerios	lets do #2
212	03:00.1	fruitloops	is there anything your not sure about?
213	04:13.3	cornflakes	no
214	04:21.8	cornflakes	i dont think so
215	04:22.9	fruitloops	me neitherr
216	05:05.1	fruitloops	anything you would like to add?
217	05:11.2	cornflakes	no i dont think so
218	05:28.5	cheerios	no its the same thing from relationships

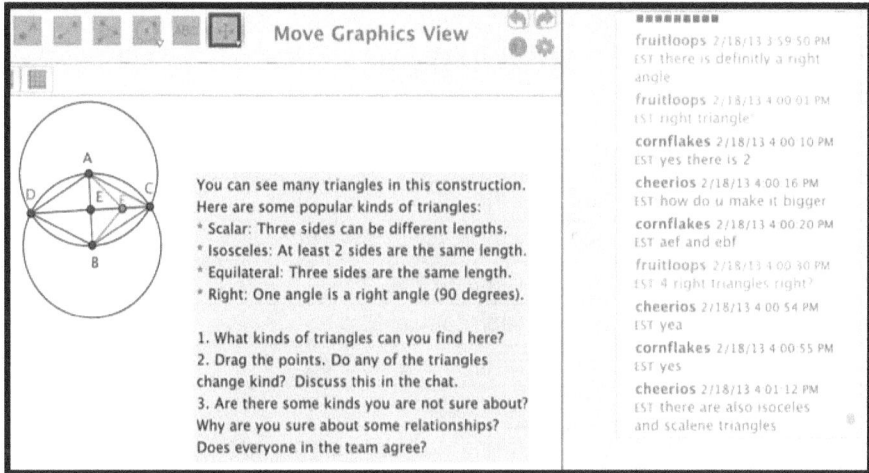

Figure 13. The Where's Waldo Tab.

Cheerios suggests, "lets do #2" (line 211). This is the interesting step in the tab. It instructs the students to drag the points and to discuss if any of the triangles change kind. If one drags point F to coincide with point C, then isosceles triangle ABF coincides with equilateral triangle ABC, suggesting that in dynamic geometry objects can change their characterizations. This could lead to an interesting discussion about how equilateral is a special case of isosceles, which is a special case of scalene (depending on exactly how one defines these categories). Ironically, it would be particularly challenging of Cheerios' tendency to classify figures based on their static appearance. However, the team barely drags the figure in this tab, not moving point F at all. No one responds to the question of step 2, even though Cheerios proposed considering it.

In response to the question of step 3, whether there is anything the team is unsure about and whether they are sure about some relationships, the team has nothing to say. They do not address issues of necessity in geometric relationships. Having missed the point of step 2, they see nothing new in this tab to discuss. With just a couple of minutes left for the session, they turn to the final tab.

Tab Exploring

After some discussion of who should take control of GeoGebra for this tab, Cheerios drags each of the triangles in the tab (**Log 19** and **Figure 14**). She rotates them and drags them larger. Then Fruitloops drags a number of them as well, apparently trying to see which can form isosceles triangles or which can match Poly1.

Log 19. Exploring triangles.

229	08:14.4	cornflakes	if the circle is black it has constraints?
230	08:17.9	fruitloops	who wants to takes turn
231	08:53.6	cheerios	u can go first
232	09:01.4	fruitloops	yeah cornflakes go first
233	09:21.4	cornflakes	im nt sure what ro do
234	09:26.7	fruitloops	me neither
235	09:42.0	fruitloops	cheerios i guess its your turn....
236	09:53.6	fruitloops	2 more mintues
237	09:53.8	fruitloops	
238	10:06.8	fruitloops	the triangles are moving
239	10:20.6	cornflakes	theyre getting bigger
240	10:24.7	fruitloops	i see what you are doing cheerios
241	10:36.3	fruitloops	can i try for a minute
242	10:39.9	cheerios	yea
243	10:40.8	cornflakes	yes
244	11:16.3	cheerios	we should dicuss about the strenghts
245	11:34.2	cheerios	constraints

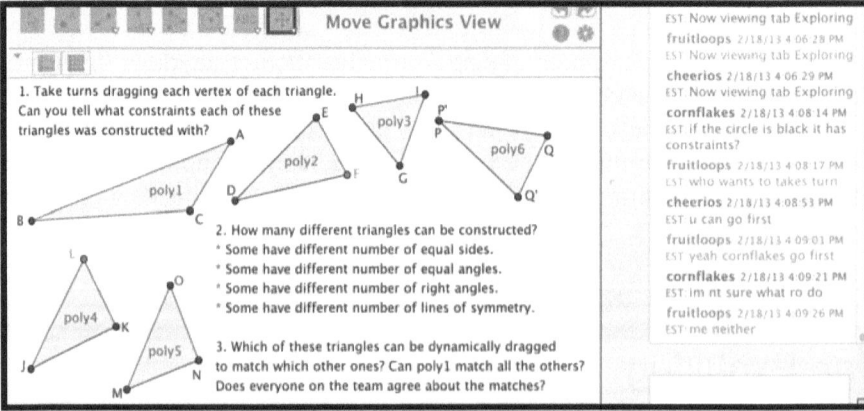

Figure 14. The Exploring Tab.

Eventually, the team realizes they should be chatting about the constraints designed into each triangle (line 244 and 245 in response to step 1). Unfortunately, their session is over. They will return to a similar task in Session 8. By then, they will be far better equipped to explore and discuss the dependencies of the various triangle constructions.

Summary of Learning in Session 2

In their second session, the team began to use the drag test and to understand its significance. The primary development during Session 2 is probably in the area of beginning to understand dynamic dragging. We have labeled a number of observed *group dragging practices* as follows:

1. *Do not drag lines to visually coincide with existing points, but use the points to construct lines between or through them.*
2. *Observe visible feedback from the software to guide dragging and construction.*
3. *Drag points to test if geometric relationships are maintained.*
4. *Drag geometric objects to observe invariances.*
5. *Drag geometric objects to vary the figures and see if relationships are always maintained.*
6. *Some points cannot be dragged or only dragged to a limited extent; they are constrained.*

This list may seem somewhat redundant. However, the drag test is central to dynamic geometry and serves multiple related purposes, involving dragging points and other geometric objects for diverse reasons.

The team's approach to working on a task like constructing an equilateral triangle is obviously mediated by their practical understanding of actions with GeoGebra tools (Carreira et al., 2016). This does not just involve construction of points, segments, circles and polygons, but also the location of these objects as they are created in relation to each other—for instance, how to construct a circle that is centered on one existing point and is constrained to pass through another existing point. In addition, tool usage involves the ability to drag geometric objects in order to establish or test various relationships among them. Dragging is a defining characteristic of dynamic geometry, and developing a command over the multiple uses of dragging in GeoGebra is essential to being able to accomplish tasks in it.

In this session, the team makes considerable progress in developing their understanding of this aspect of dynamic geometry by adopting important group dragging practices. However, their understanding of dragging and construction in dynamic geometry is still quite weak. For instance, it took them quite a while to make initial progress on constructing the equilateral triangle. In several instances, the team has not fully enacted the lessons intended by the instructions.

The team's discourse is starting to be more productive and they explicitly discuss ideas about dependency. In dynamic geometry, dependency relationships are often best determined by the drag test. While the team has begun to develop practices related to the drag test, they have not adopted its consistent use. We shall follow how this continues to evolve in their future sessions.

At the end of the Cereal Team's second session, we see some ways in which the team continues to form into an effective group for exploring dynamic geometry:

i. It solidifies its basic *collaboration practices*, taking turns more equally on the GeoGebra construction and discussing what they are doing in the chat. At the start of the session, they do not communicate about their GeoGebra actions and they flounder. Later they discuss in the chat and are much more productive.

ii. The team increases its *productive mathematical discourse*, discussing their construction work together. They also begin considering in the chat what relationships hold for a figure "always." They start to use the term "constraint," although they still have only a vague notion of the word's meaning and proper application.

iii. Each of the students engages in *constructing and dragging GeoGebra* objects. They use the drag test more and they all note its consequences.

We can begin to track the team's increasing fluency with *identifying and constructing dynamic-geometric dependencies*:

a. The team does more *dynamic dragging* of points. Although they have begun to use dragging to investigate figures, they have not adopted the drag test as a regular practice for making sure that their constructions are dynamically valid. They also often continue to rely on visual appearance rather than on behavior under dragging to characterize and understand figures. They do not always use dragging to vary a figure beyond its initial static appearance—in order to determine what characteristics a constructed figure necessarily has as a dynamic figure.

b. They eventually succeed in *dynamic construction* of the equilateral triangle. This increases their skill level in constructing figures that include circles connected to existing points and segments.

c. However, the team does not begin to design *dynamic dependency* into GeoGebra constructions on their own, without step-by-step instructions.

We can draw some suggestions for re-design of the topics based on the behavior of the team in Session 2. How can the team avoid floundering for a half an hour on the construction of the equilateral triangle? How can they more clearly learn about how to do GeoGebra constructions so that the desired relationships are captured by the software?

- Clearly, the wording of the construction can be elaborated so that it specifies more explicitly where to click, etc.

- Perhaps some preliminary construction exercises should be included first, such as constructing a circle using the endpoints of a circle and doing a drag test to see that the relationships hold under some construction methods and not under others.

- The individual students could even be given some opportunities to explore or play with GeoGebra on their own before working together, to avoid spending excessive group time floundering and experimenting.

- The team observed the importance of the drag test because it was prompted for in an appropriate context of construction. How can this lesson be emphasized so that the team will start to use the drag test more regularly to check the effectiveness of their constructions? It is not just a matter of dragging any point, but of systematically checking the validity of key relationships that were intended.

- It should be mentioned that dragging should be vigorous, so that the figures are changed to vary through all their possible appearances and special cases.

- It should be emphasized that the figures should be considered dynamic, with relationships that are maintained under dragging. This should be contrasted with temporary static visual appearances.

- Students should be encouraged to display for their teammates using GeoGebra the answers they develop to discussion prompts.

Session 3: The Team Develops Construction Practices

Topic 3 is designed to build on the previous topic's experience of constructing an equilateral triangle to provide further experience in constructing dynamic figures, such as perpendicular bisectors and parallel lines. The first tab shows how the complex figure from the previous topic involved a perpendicular bisector. It then presents the challenge of constructing a perpendicular to a segment through a given point on that segment. The second tab is for defining a custom tool to make perpendicular lines. This can involve variations depending on whether the perpendicular is to go through a given point on or off the original segment—and thereby provides opportunities for open-ended exploration.

Before working on this topic online, the students participated in a class presentation by their teacher. The teacher displayed the last tab of Topic 2, which most students had not had time to work on. She discussed the notions of free, constrained and dependent points. She also distributed paper copies of Topic 3, allowing the students to create more work space by removing the instructions from their GeoGebra tabs and still be able to follow the specified steps.

Visual Drawings and Theoretical Constructions

In their work on Topic 3, the Cereal Team begins to learn about the difference between visual "drawings" and theoretical dynamic-geometry "figures" or "constructions." This distinction lies at the heart of dynamic geometry and is related to various cognitive considerations.

As mentioned earlier, a popular theory in mathematics education is that of van Hiele levels (van Hiele, 1986; 1999). This theory proposes that students must progress through a series of levels to understand mathematics the way that mathematicians do. Children start at the level of visual appearances, having been acculturated to recognize the visual shapes of circles, triangles, squares, perpendicular lines, etc. in terms of prototype images (Lakoff, 1987; Lakoff & Núñez, 2000; Rosch, 1973). They must undergo a cognitive development to reach the next level, understanding geometric figures in terms of relationships among their parts, such as equal sides forming a square. In this session, the Cereal Team starts to make such a transition.

The contrast of visual appearance to geometric relationships has been conceived in other theories as well. Vygotsky (1934/1986) stressed that education should transform everyday conceptualizations into scientific forms. Current programs of educational reform call for 21st Century skills, including fluency with mathematical formulations of relationships. Such mathematical thinking is also central to the emphasis on science, technology, engineering and mathematics (STEM) skills. Studies of other VMT research (Çakir & Stahl, 2013; Çakir et al., 2009) has distinguished the roles of verbal (discourse), visual (graphical) and symbolic (algebraic) communication within math problem solving, and how they support each other, often building sequentially from an initial naïve narrative understanding of a problem to a mathematical formulation of its solution. Frequently, numeric measurement supplements the visual aspects, for instance with measuring side lengths to confirm visual judgments—rather than arguing from relationships among the constructed components.

In published research on dynamic geometry, there is an important distinction drawn between a *drawing* and a *construction*. "Drawing" refers to the juxtaposition of geometrical objects that look like some intended figure (Hoyles & Jones, 1998). "Construction," however, depends on creating relationships—in other words dependencies—among the elements of a figure. In dynamic geometry in particular, once relationships are defined and constructed accordingly, the figure maintains these theoretical relationships even under dragging. The transition from visual to formal mathematics, nonetheless, has been found to be neither straightforward nor easy for students working with dynamic geometry (Jones, 2000). Students often think that it is possible to construct a geometric figure based on visual cues (Laborde, 2004), as we shall see in the beginning of this session.

One can also make the distinction dialogically, between two different mathematical discourses (Sfard, 2008a). Sfard (2008b) analyzes thinking/communicating in mathematics in terms of meta-rules: actions of participants that relate to the production and substantiation of object-level rules. Sets of meta-rules that describe a patterned discursive action are named *routines*, since they are repeated in certain types of situations. In reviewing the logs of Session 3, we will observe two contrasting production routines: (i) visual placement by drawing or dragging and (ii) construction by creating objects with dependencies between them. The verification of the team's perpendicularity routines are sets of procedures describing the repetitive actions they take in substantiating whether a newly produced line is in fact perpendicular to a given line. We will observe two contrasting verification procedures: (i) visual judgment or measurement, and (ii) use of theoretical geometry knowledge to justify proposed solutions.

Session 3 has been analyzed in (Öner & Stahl, 2015a; 2015b) as consisting of a series of five interactional episodes, in which the team moves incrementally from a purely *visual* approach to one that begins to incorporate a sense of *theoretical* construction of dependencies. The session is divided into the following episodes:

a) Reconstruct the example figure (see **Log 20** below).
b) Draw a perpendicular-looking line (**Log 21**).
c) Use the example figure as a guide (**Log 22**).
d) Use circles without dependencies (**Log 23**).
e) Construct dependencies (**Log 24**).

In this chapter, we will follow that analysis, conceiving it in terms of *group construction practices* rather than Sfard's and Öner's closely related construct of mathematical-discourse routines. We shall see how the team begins to make a transition from visual drawing to theoretical construction through the adoption of a series of several group construction practices.

Tab Bisector

From the very start, this session is about learning to use GeoGebra's construction tools to accomplish tasks, solve geometric problems and answer related questions. Fruitloops starts this session by following the instructions for constructing a perpendicular bisector, as shown in **Figure 15**. She makes a pair of points, I and J, several times and then asks "how do i make the line segment?" (line 17 in **Log 20**). The situation here is different from when Fruitloops asked a similar question in Session 1 (line 80 in **Log 5**) and Session 2 (line 94 in **Log 14**). In the GeoGebra tabs for the previous topics, the segment tool was visible in the tool bar. In Session 3, now the line tool is visible and the segment tool has to be pulled down from behind it. Cornflakes responds, indicating that the segment tool is next to the circle tool in the toolbar. Then Fruitloops selects the segment tool and connects her two points with segment IJ.

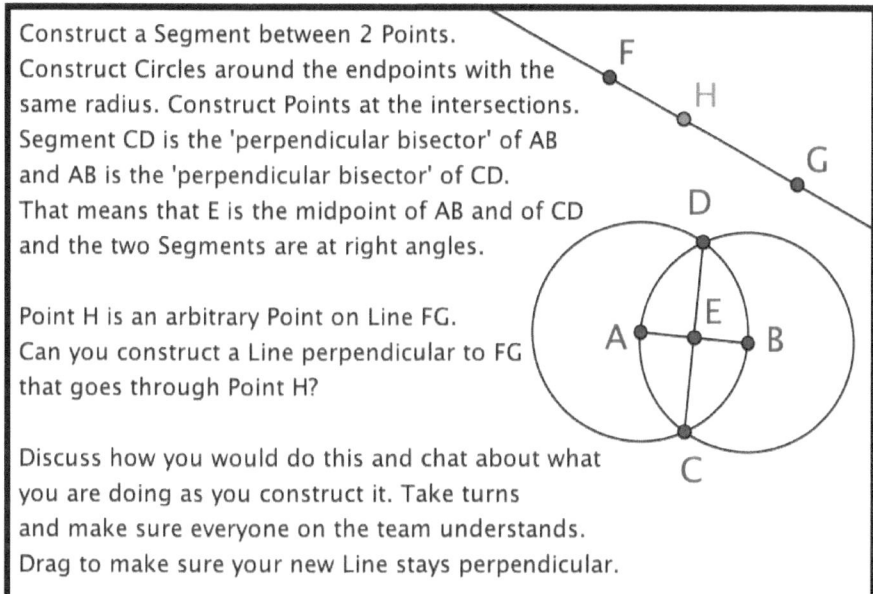

Construct a Segment between 2 Points.
Construct Circles around the endpoints with the
same radius. Construct Points at the intersections.
Segment CD is the 'perpendicular bisector' of AB
and AB is the 'perpendicular bisector' of CD.
That means that E is the midpoint of AB and of CD
and the two Segments are at right angles.

Point H is an arbitrary Point on Line FG.
Can you construct a Line perpendicular to FG
that goes through Point H?

Discuss how you would do this and chat about what
you are doing as you construct it. Take turns
and make sure everyone on the team understands.
Drag to make sure your new Line stays perpendicular.

Figure 15. The Bisector Tab.

Log 20. Reconstruct the example figure.

17	33:03.9	fruitloops	how do i make the line segment?
18	33:08.0	cheerios	do u need help
19	33:26.1	cornflakes	its by the circle thingy
20	33:38.1	fruitloops	got it thanks
21	34:06.5	cornflakes	no problemo
22	35:54.1	fruitloops	i did it
23	36:02.0	cheerios	good job my peer
24	36:15.6	fruitloops	someone else want to continue?
25	36:14.4	cornflakes	Nice
26	36:23.6	fruitloops	thankyou thankyou
27	36:32.5	cheerios	release control
28	37:40.4	fruitloops	so now you need to construck points at the intersection
29	38:12.1	fruitloops	no you dont make a line you make a line segment
30	39:29.9	cheerios	i just made the intersecting line and point in the middle
31	38:35.1	fruitloops	good!!

| 32 | 39:20.4 | fruitloops | so continue |
| 33 | 39:40.0 | cheerios | it made a perpindicular line |

Fruitloops next creates a circle centered on point I and an overlapping circle centered on point J. However, she creates each of these circles by placing a point approximately where she thinks the circumference should go (just as Cornflakes started to do in the previous topic, which Fruitloops questioned in line 90 of **Log 13**). This creates a figure that looks similar to the visual appearance of the example figure, but does not have the dynamic relationships that were in the example. In particular, the instructions mention that the circles should have "the same radius." Constructing the circles to have the identical radius is accomplished in the example figure by using segment AB as the radius of each circle: The circle centered on point A is defined by and goes through point B and the circle centered on point B is defined by and goes through point A. Fruitloops has not done that. She has created circles that have radii that look about the same length. She is working in the visual paradigm. She has not acted as though dependencies can be established as the result of how geometric elements like circles are constructed in GeoGebra.

After Fruitloops says, "got it thanks" (line 20) for the help in finding the segment tool icon next to the circle one, she must see that her circles are the wrong size and she deletes them and tries a couple more circles. She creates a series of six circles of various sizes to explore how the circle tool works. Although she seemed to know how to construct circles using a segment as a common radius in Topic 2, she had not done the construction herself there but had guided the others. Now, in Topic 3, may be the first time that Fruitloops actually constructs a circle herself in GeoGebra, and she needs to play around with it to see how the circle tool works. Finally, she constructs the two intersecting circles, both with radius IJ and says, "i did it" (line 22). She then releases control to Cheerios.

> *Group construction practice #1: Reproduce a figure by following instruction steps.*

Cheerios completes the first part of the instructions by constructing points K and L at the intersections of the circles. First, she creates a line through K and L, using the line tool. Then, she erases it and creates it again. Like Fruitloops, she must access the segment tool from behind the line tool. They are now aware of the distinction of lines (which proceed indefinitely past the defining points) and segments (which end at their endpoints). Perhaps they became aware of this by seeing the two different tools and then checking the wording of the instructions—and then shared the distinction within the team. With guidance from Fruitloops, Cheerios replaces the line with a segment and marks point M at the intersection of the two segments, as seen in **Figure 16**. Cheerios calls point M the "point in the middle" (line 30) and notes, "it made a perpindicular line" (line 33).

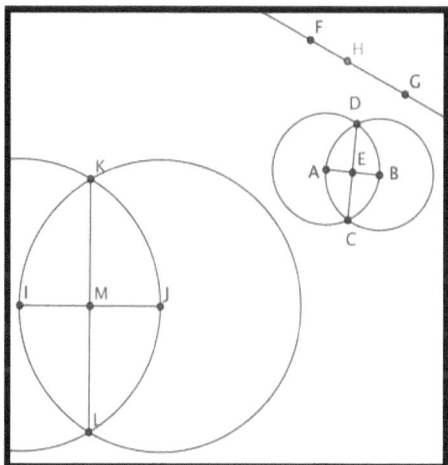

Figure 16. Constructing the midpoint.

At this point in the session, both Fruitloops and Cheerios have succeeded in constructing the specified figure. They have had to engage in some hands-on experimentation or trial-and-error. They have also had to pay careful attention to the exact wording of the task instructions and the details of the example figure—such as the distinction between a line and a segment or the requirement that the circles have the same radius. Cornflakes, who is generally more engaged in trying constructions and consequently more skilled at construction, has been quiet while the others catch up to her skill level. For the more challenging part of the tab, Cornflakes takes GeoGebra control.

The instructions in the GeoGebra tab next ask the students to construct a line perpendicular to line FG, which goes through a point H that already exists on line FG. (There is a relatively obscure trick in this. First, the given point must become a midpoint between two points; then those two points can be used as the centers of intersecting circles, with the line between their intersections passing through the original point, as shown in the next tab. The students never discover this trick.)

Log 21. Draw a perpendicular-looking line.

| 34 | 40:27.5 | fruitloops | okay cornflakes go next |
| 35 | 41:11.5 | cornflakes | what are you supposed to do? |

36	41:42.6	fruitloops	just follow the instructions
37	43:48.5	fruitloops	were we supposed to just use the line that was already there?
38	44:10.2	cornflakes	i think so
39	44:44.2	fruitloops	perpindicular no intersecting
40	44:46.1	fruitloops	not*
41	48:09.7	fruitloops	sorry i did it by accident
42	48:23.5	cheerios	its fine :) my dear peer
43	48:38.3	fruitloops	can you remake it
44	48:52.7	fruitloops	why did you make point o and q
45	48:55.0	cornflakes	its alright
46	49:09.5	cheerios	is the line ok
47	49:16.0	cornflakes	i didnt make point o and q
48	49:23.0	fruitloops	its not perpinicuklar
49	50:57.7	fruitloops	i think thats good
50	50:59.8	cheerios	turn line fhg so its easier make it horizontal
51	52:54.4	fruitloops	hey
52	54:06.9	fruitloops	which point did you move to get the line like that
53	54:07.5	cornflakes	now construct the line
54	55:10.7	cornflakes	thats good
55	55:30.5	fruitloops	i think its perpendicular cause they are all 90 degree angles

Each of the three team members tries repeatedly to create a line that goes through point H and that looks like it is perpendicular to line FG. They do this by using H as one of the points that defines the line and then by creating a second point and dragging it until it makes their new line look visually perpendicular to the existing line FG. This is a clear example of the visual misconception approach. The team is now using this as a group practice.

> *Group construction practice #2: Draw a figure by dragging objects to appear right.*

After they have each tried without satisfying everyone else that their line is perpendicular, Cheerios suggests that they "turn line fhg so its easier make it horizontal" (line 49). While orientation has no effect on the validity of a dynamic-geometry construction, the prototypical visual image of perpendicular lines has them going horizontal and vertical respectively. Cheerios apparently wants to re-orient their figure to make it easier to mentally compare its visual appearance to the prototype image of perpendicular lines. Cheerios rotates the line and draws a vertical looking line that appears to be perpendicular to the now horizontal line FG. Fruitloops makes a judgment based on visual appearance that the angles formed by Cheerios' crossed lines "are all 90 degree angles" (line 55). This is the measurement misconception approach, which the group is now following.

> *Group construction practice #3: Draw a figure by dragging objects and then measure to check.*

Cornflakes next drags the example figure with AB perpendicular to CD and tries to place it over line FG and the new line through H. That will use the example figure with its perpendicular to test the perpendicularity of the students' line (rather like using the example figure as a protractor). Fruitloops assists in rotating the example figure to line it up with the team's lines and show that their lines are close to perpendicular. Cornflakes explains what they have done in line 57 of **Log 22**. She then suggests constructing their new perpendicular to FG by constructing it through the overlaid segment AB. This is a clever and creative approach, but is not based on anything the instructional guidance has ever suggested. It is another form of measurement. It is not the dynamic-geometry way of doing things and would not result in precise or dynamically valid constructions. In lines 59 and 60, Fruitloops objects to Cornflakes' proposals about how to proceed. This is a variation of the measurement practice.

> *Group construction practice #4: Draw a figure by dragging objects to align with a standard.*

Log 22. Use the example figure as a guide.

56	56:28.6	cornflakes	yes
57	57:05.2	cornflakes	so after construting the line we put the circle on top
58	57:56.8	cornflakes	so put the line thru the line on the circle
59	58:18.5	fruitloops	i dont know what i am doing help
60	58:24.8	fruitloops	sonmeone else take control

A half minute later, Fruitloops brings the group back to the approach of constructing with circles: "i think you need to make the circles first" (line 62 in **Log 23**). This is the first time that the group seems to connect their previous work constructing an equilateral triangle—or even their previous work in this tab re-constructing the example perpendicular—to their current task of constructing a perpendicular through point H on line FG. Until now, they have approached the task through attempts to match the visual appearance of perpendicularity. Now they switch to taking a dynamic-construction approach based on their recent construction of the equilateral triangle and the perpendicular bisector.

Note that the curriculum presented in Sessions 1 and 2 had introduced the tools and procedures needed for Session 3. It involved the tools to construct and drag points, segments, lines, circles and polygons. It also involved procedures for constructing figures out of these objects, in which the objects are related in specific ways. In particular, it showed that the construction of the equilateral triangle using circles led to the construction of a perpendicular bisector in the Where's Waldo Tab of Topic 2. So the students had already had some experience with these tools, procedures and related conceptualizations. However, for the first 26 minutes of session 3, the team had not applied any of this recent experience to their current problem.

The problem-solving move of applying previous solutions to new problems is itself a practice that must be adopted. This is an important lesson in mathematical cognition. For instance, in his well-known book on problem solving, Polya (1945/1973) repeatedly recommends finding a related problem that one already knows how to solve.

> *Group construction practice #5: Use previous construction practices to solve new problems.*

The curriculum had been designed to build sequentially on construction practices, particularly those involving dependencies. In particular, the construction of simple geometric objects—including points at the intersection of lines and circles—was followed by the construction of an equilateral triangle—with its dependency of side lengths upon circle radii. These provide the tools needed for the construction of perpendicularity, as shown in the Where's Waldo Tab. This parallels the application by Euclid (300 BCE) of the equilateral triangle's dependency relationships to many other geometric constructions. However, it requires the practice of recognizing related problems and adapting their solution to new problems. The students in the Cereal team were so used to looking at geometric figures as visual shapes, that it was only after breakdowns, running into deadends with that approach during the first half of their session, that they began to adopt the approach of the equilateral-triangle construction.

This is a key turning point in the team's work. They erase the figures they had drawn using just visual criteria and then re-do their work using a construction approach. They struggle to construct the figure of perpendicular lines using circles and related procedures from their construction of the equilateral triangle in the previous session and from their re-construction of the example figure (see **Log 23**).

Log 23. Use circles without dependencies.

61	58:35.8	cheerios	make the line first
62	58:51.2	fruitloops	i think you need to make the circles first
63	59:19.0	cornflakes	put point m on tp of h
64	02:26.9	fruitloops	the line isnt going through part h
65	02:39.5	cornflakes	bisection is a division of something into two equal parts
66	04:58.2	fruitloops	we didnt put a point between the circles so the libne isnt perpendicular
67	05:03.8	fruitloops	line*
68	05:20.6	fruitloops	the part where the circles intersect
69	05:19.4	cheerios	oh i see now
70	05:34.8	fruitloops	look at the examples and youll see
71	05:46.9	cornflakes	ok i see
72	05:51.8	cheerios	r u fixing it
73	05:54.7	fruitloops	do you want to do it?
74	06:02.0	cornflakes	so we have to put a poijt bewtween the circles

Fruitloops constructs circles that are centered on F and G. To follow the actual construction steps taken by the students, we must analyze their GeoGebra actions in the VMT Replayer. Although the circles look like they are going through points G and F, they are actually defined by new circumference points—just as Cornflakes had originally done in the previous session. Fruitloops created the circle around F by clicking first on point F and then clicking on a new point and dragging the new point until the circle seemed to pass through G before releasing it to define the circle. Similarly, for the circle around G she clicked first on G and then on a new point and, before releasing the cursor to form the circle, she dragged the new point until the circle appeared to pass through point F. This procedure made it appear visually that both circles had radii of FG.

> *Group construction practice #5: Construct equal lengths using radii of circles.*

However, the segment FG was not actually used to define the circles. If segment FG ever changed, the circles would not dynamically change accordingly. Thus, the construction did not have the necessary dependencies. The students have learned that in creating a circle, one can define new points for the center and circumference and drag them to create the circle where one wants it. This is an affordance of the GeoGebra circle tool, but not the one needed to establish effective dynamic relationships of dependency on existing points and lines.

Cheerios then takes control and constructs a line that appears to go through the intersections of the circles, although it is actually defined by two new points created above and below the visible construction area. Although they are mimicking the construction procedure for a perpendicular, both Fruitloops and Cheerios are locating free points at locations that create circles and lines that look like they have the relevant geometric relationships, but do not actually embody the necessary construction dependencies.

Cheerios sees that her new line misses point H. The instructions had called for it to intersect line GF at point H. In response, Cheerios simply moves point H over to where the new line intersected FG. She also tries moving the line back and forth to intersect H. However, nothing seems to work and Fruitloops complains, "the line isnt going through part h" (line 64).

The team realizes that the line formed by the intersections of the circles is a bisector of the distance between the centers of the circles (Cornflakes in line 65), and that point H is not at that midpoint the way it was in the example figure (Fruitloops in lines 66-70). To fix this problem, Cornflakes starts to drag the circles. This reveals that they were not constructed with FG as their radius; they do not hold up dynamically under this drag test. Cornflakes drags points F, H and G so that eventually H looks like it is on the line that was supposed to be perpendicular. She uses the ability to drag points and lines in order to establish visual appearances, not to establish or test for dynamic dependencies. Because the team's figure was not constructed with the proper dynamic dependencies, it is now a mess. The team has to start over, trying to avoid the problems they ran into. They learn from the feedback of their construction in GeoGebra that the way they had proceeded was invalid.

Now Fruitloops constructs two circles and properly uses segment FG as their radii (line 79 in **Log 24**). She uses point G to specify the center of the first circle and point F to define its circumference. Then she uses point F to specify the center of the other circle and point G to define its circumference. Cheerios and Cornflakes agree that this construction creates equal radii, whose length is given by segment FG (lines 80 and 82). The team has finally constructed the circles in a dynamically valid way, although they have not checked that with a drag test.

> *Group construction practice #7: Construct an object using existing points to define the object by those points.*

Log 24. Construct dependencies.

79	08:23.3	fruitloops	so i madfe two circles that intersect and the radius is the same in both circles right?
80	08:41.9	cheerios	yea they are the same
81	08:55.1	fruitloops	and segment fg is the radius
82	08:58.4	cornflakes	yes
83	09:04.1	cheerios	now we have to make another line
84	09:14.8	fruitloops	yeah someone else can do that
85	11:09.8	fruitloops	you make the points go through qr and then you move h ontop of the line
86	13:08.4	fruitloops	i think i did it finallyu
87	13:49.1	cornflakes	the klines bisec the circle
88	14:15.3	cornflakes	*the lines bisect the circle

Fruitloops turns control over to Cornflakes to connect the intersection points Q and R with the perpendicular line. Cornflakes constructs the connecting line and then deletes everything and constructs the circles again, using the procedure Fruitloops had used, thereby demonstrating her understanding and acceptance of it. Fruitloops takes control and constructs line QR through the intersections of the circles again.

As can be seen in **Figure 17**, line QR does not pass through point H. To fix this, Fruitloops simply drags point H over so that it looks like it is on line QR. As she says, "you make the points go through qr and then you move h ontop of the line" (line 85 in **Log 24**). After having done the dynamic construction, she adjusts it non-dynamically to make things appear visually correct.

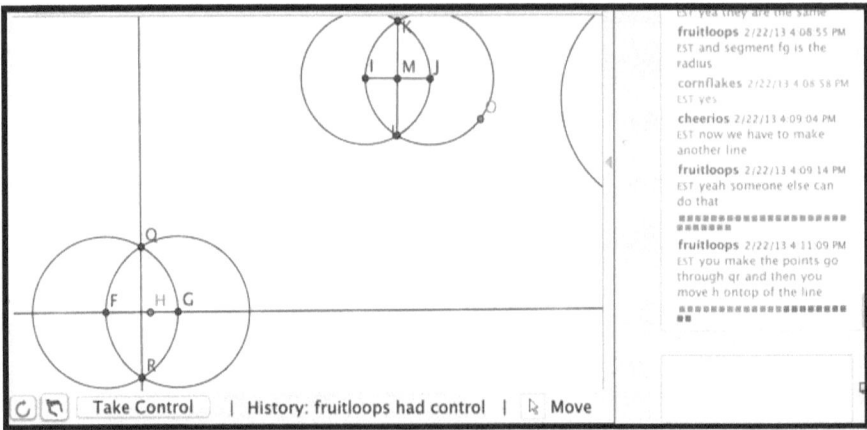

Figure 17. Adjusting the midpoint.

It may seem that dragging point H is legitimate in dynamic geometry, where points can be dragged to new locations. However, this is a subtle misconception on the part of the students. Point H was given as part of the problem. While in this particular situation, it can be moved, in other construction settings it might be constrained as part of another figure or as a point on a line, etc. The task in this tab is designed to call for a construction procedure to deal with a point that is not midpoint between the endpoints of the given base segment. To drag the point back to a midpoint obviates the intended problem and is not in general possible. Perhaps the topic should have been designed or stated to disallow such dragging.

Fruitloops announces that they have finally succeeded in their task: "i think i did it finallyu" (line 86). Cornflakes notes a consequence based on their having followed the perpendicular-bisector construction procedure: that line QR bisects segment FG (line 87). However, Fruitloops follows this by questioning how they know that they have actually constructed a perpendicular as they were tasked to do: "but how do we know for sure that the line is perpinmdicular" (line 89 in **Log 25**).

Log 25. Discuss why the figure is perpendicular.

89	14:29.8	fruitloops	but how do we know for sure that the line is perpinmdicular
90	14:39.6	cheerios	im not sure
91	14:42.1	cornflakes	there 90 degree angles
92	14:45.4	cheerios	do u cornflakes
93	14:59.4	fruitloops	but you cant really prove that by looking at it

94	15:06.8	cornflakes	they intersect throught the points that go through the circle
95	15:17.7	fruitloops	it has to do with the perpendicular bisector
96	15:19.8	cornflakes	they"bisect" it
97	15:31.2	fruitloops	and the circles
98	15:37.2	cheerios	oh i see

Cheerios passes on trying to answer this. Cornflakes makes a first attempt to say how they know that their new line is perpendicular. She says that it forms a 90-degree angle (line 91). Fruitloops responds to this with the intriguing statement, "but you cant really prove that by looking at it" (line 93). This explicitly mentions the issue of proving one's claims. Fruitloops contrasts proof with what one can see by looking. While it may look like the line is perpendicular, one cannot tell visually that it is exactly 90 degrees. Then, both Cornflakes (in lines 94 and 96) and Fruitloops (in lines 95 and 97) connect the proof of perpendicularity to the construction process. While they do not explicitly state that the construction process was designed to produce a perpendicular line, they indicate that a potential proof "has to do" with the construction procedure. This marks a glimmering recognition of the connection of construction procedures to rigorous explanations. Cheerios aligns with her partners. They then move on to the next tab.

> *Group construction practice #8: Discuss geometric relationships as results of the construction process.*

This may be the first time that the team shifted from a purely visual concern with the graphical appearance of relationships like perpendicularity to a mathematical consideration of construction elements that could be relevant to a logical proof. Fruitloops raises the issue of provability and both she and Cornflakes relate it to the kinds of construction relationships that they have been involved with in this and the previous topic. They do not go beyond a vague reference to such factors.

Before the team moves on, Cheerios follows the final instruction on the tab: "Drag to make sure your new Line stays perpendicular." She vigorously drags their construction around. It maintains its structure, although no one comments on this drag test.

> *Group construction practice #9:*
> *Check a construction by dragging its*
> *points to test if relationships remain*
> *invariant.*

Tab Perpendicular

The next tab involves defining a "custom tool" in GeoGebra to create perpendicular lines, given two points defining a base line. The tab also illustrates with its figure a proper solution to constructing a perpendicular line FG through a point C that is on a line AB, but is not at the midpoint of segment AB (see **Figure 18**).

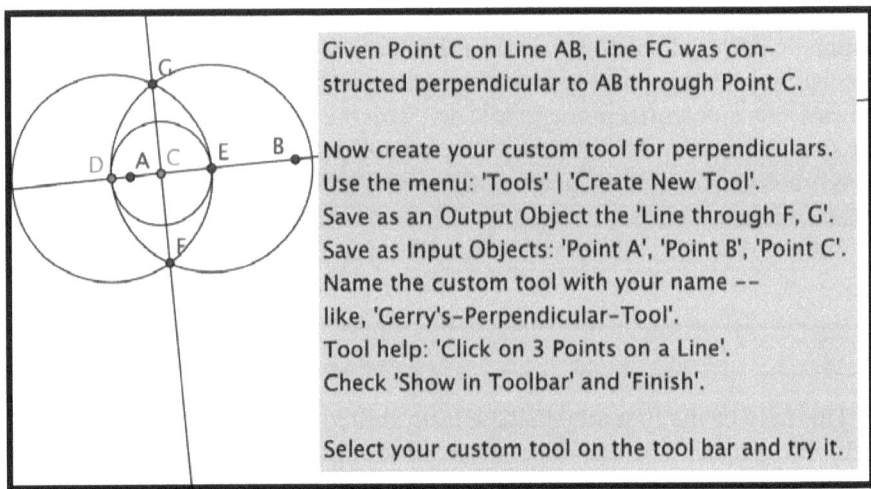

Given Point C on Line AB, Line FG was constructed perpendicular to AB through Point C.

Now create your custom tool for perpendiculars.
Use the menu: 'Tools' | 'Create New Tool'.
Save as an Output Object the 'Line through F, G'.
Save as Input Objects: 'Point A', 'Point B', 'Point C'.
Name the custom tool with your name --
like, 'Gerry's-Perpendicular-Tool'.
Tool help: 'Click on 3 Points on a Line'.
Check 'Show in Toolbar' and 'Finish'.

Select your custom tool on the tool bar and try it.

Figure 18. The Perpendicular Tab.

The team does not discuss the figure or even what they are doing. Cornflakes defines a custom tool. She uses the existing figure as input to the custom tool interface, rather than re-creating her own figure based on the example. Thus, the team did not benefit from the intended lesson in constructing a perpendicular through a point on the base line, but not at its midpoint. No one displays an ability to view the example figure as a solution to the construction that they just worked so hard on. They have not developed the ability to see figures as informative visualizations of interesting construction procedures, let alone as proofs of relationships. Nor have they discussed how the sequences of tasks they are given in the tabs and topics are related to each other.

Cornflakes succeeds in creating a custom tool, which becomes available to the whole team. Cornflakes uses her custom tool to successfully create a line that is perpendicular to a line that would pass through the two points selected with the tool. However, she does not believe that her tool is working properly (line 105 in **Log 26**). It may be that Cornflakes and the others do not see that the custom tool worked because the line connecting the base points (IH or HJ in the two uses of the custom tool) is not shown. Thus, it is hard to see that the new line that appears looks perpendicular to a base line (HJ in **Figure 19**). No one else in the team tries to use the tool. There is no discussion. The team runs out of time before getting to the Parallel Tab.

Log 26. Defining the team's custom perpendicular tool.

102	17:03.9	fruitloops	someone take control and do step 1
103	26:06.6	fruitloops	try out your tool bar
104	26:13.7	fruitloops	your tool*
105	28:33.9	cornflakes	my tool isnt working
106	28:39.5	cheerios	try doing it again
107	28:46.6	cornflakes	i did
108	28:52.9	cornflakes	try making a tool
109	30:06.4	fruitloops	but didnt you make one already

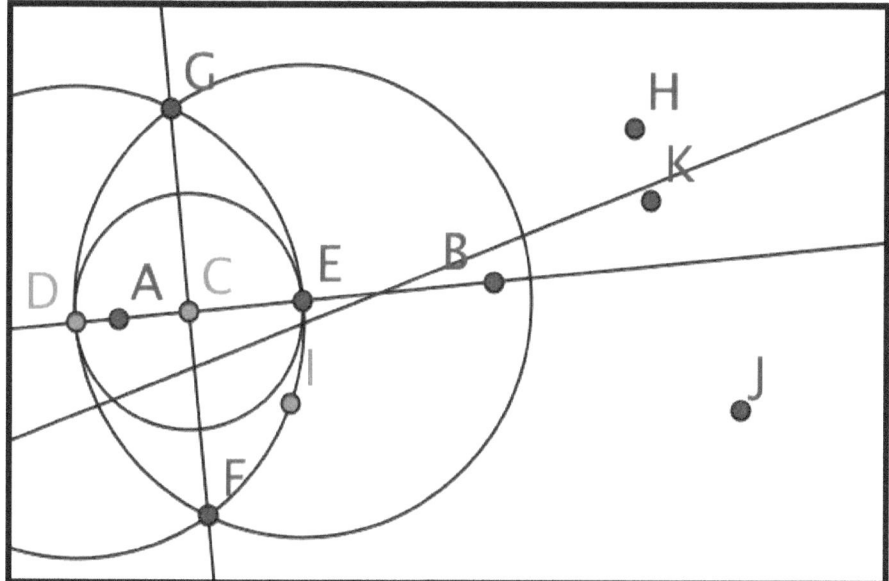

Figure 19. A perpendicular to HJ.

Summary of Learning in Session 3

In their third session, the team's major change involves their construction work. Their understanding of the dynamic-geometry approach to building dependencies into figures through specific kinds of construction is still quite fragile at the end, but they have begun to experience the difference between drawing based on visual inspection and construction based on geometric features. They spend considerable time trying to make figures look right visually, before figuring out how to construct them with the dynamic relationships built in. We have labeled their sequence of observed *group construction practices* as follows:

1. *Reproduce a figure by following instruction steps.*
2. *Draw a figure by dragging objects to appear right.*
3. *Draw a figure by dragging objects and then measure to check.*
4. *Draw a figure by dragging objects to align with a standard.*
5. *Construct equal lengths using radii of circles.*
6. *Use previous construction practices to solve new problems.*
7. *Construct an object using existing points to define the object by those points.*
8. *Discuss geometric relationships as results of the construction process.*
9. *Check a construction by dragging its points to test if relationships remain invariant.*

In working on Topic 3, the team moved from attempts to *draw* perpendicular figures (practices 1, 2, 3, 4) to efforts to *construct* them (practices 5, 6, 7, 8, 9). It thus facilitated the team's movement from one van Hiele level toward the next one: from *visual* to *theoretical*.

While many researchers accept something like van Hiele's claim that students must advance through qualitative stages of manipulating and conceptualizing geometric figures, not much is known about *how* students actually accomplish this developmental change. In our analysis of the Cereal Team's work in this chapter, we have identified nine group construction practices that the team has adopted, at least tentatively or temporarily. It is precisely through such adoption of group practices that the team developed. Of course, this does not mean that every group of students has to adopt exactly this sequence of practices to move toward the theoretical level of geometric cognition. However, this analysis provides an example of how it can actually be done. The analysis secondarily provides insight into the circumstances— such as the wording of the tasks—that did or did not promote developmental progress by the team.

The curriculum seems to have played a major role in moving the team from a visual-shapes to a construction-based approach. Not only do the instructions in Topic 3 prompt considerations of dependency relationships in the construction of figures, but the sequencing of Topic 3 after Topic 2—with its highlighting of the perpendicular bisector as implicit within the equilateral-triangle construction—eventually led to the team adapting the earlier construction procedures to the new problem. This was itself an important development of the team's mathematical group cognition (practice 6).

The students focus on a series of three concepts in their discourse in Session 3: perpendiculars, arbitrary points and bisection. These are all prompted by the text, graphics and tasks of the topic. They are inter-related, but they provide distinct themes of the student discourse. They are focal points within a rich fabric of concepts, images and experiences, which stimulate productive mathematical discourse among the students. Perpendicularity is discussed as a visual feature of two lines; prototypically, one line is horizontal and its perpendicular is vertical. But perpendicularity also involves right angles and can be measured by aligning a model of perpendicular lines. There is also an emergent sense in which the perpendicularity of two lines can result from the procedure of their construction. The arbitrary point H on the given line FG is not treated as a particular given point, but as one that can be relocated anywhere on FG. The act of bisection is sometimes seen as key, but it is not analyzed in terms of its construction details, involving equal lengths. The inter-related senses of these concepts and the multiplicity of their applications within the student discourse contribute to a deepening sense of mathematical relationships of perpendicularity.

In general, Topic 3 brings together a number of ideas and geometric themes from the math community, which the individual students respond to in their group interaction. The students have observed and experimented with the equilateral-triangle construction and the related perpendicular-bisector construction. They have responded at length to the challenge of constructing a perpendicular line through an arbitrary point on a given line. Considerable work using visual-appearance-based approaches eventually evolved into construction approaches. The construction approaches were associated loosely with the possibility of providing proof.

During this session, we see continuing progress in all dimensions of the development of geometric group cognition:

i.   The team is using its basic *collaboration practices*, especially taking turns with construction and discussing the conclusions of their work. However, there are still long periods when there is little or no discourse—often corresponding to periods of individuals experimenting with the construction tools.

ii.   The team shows some *productive mathematical discourse*, especially about explaining relationships, such as perpendicularity.

iii.   Each of the students engages in *constructing and dragging* objects with the GeoGebra tools. Both Cheerios and Fruitloops take control more often now and make substantial contributions to the constructing in GeoGebra. However, the students do not yet systematically use dragging to explore relationships in figures, despite having adopted this as a group practice that they can use at any time. Dragging is used more within their drawing approach than as a test within a construction approach.

The team is becoming more comfortable engaging in dynamic-geometry activities to *identify and construct dynamic-geometric dependencies*:

a.   The students use *dynamic dragging* of points, although it is often just to adjust appearances of figures rather than to investigate the figure's dependencies. They have not adopted the drag test as a regular practice for testing the validity of their constructions.

b.   The team has succeeded in using the *dynamic construction* procedure from the equilateral triangle for constructing perpendiculars, although they did not immediately follow this approach.

c.   The use of the equilateral-triangle construction procedure allows the team to establish *dynamic dependencies* in GeoGebra constructions, but the team does not clearly articulate an understanding of this.

The tasks of Topic 3 seem to be useful in guiding the students from their orientation to visual appearances toward an understanding based more on structural relationships established through construction and tested through dragging. The shift from the base van Hiele level of visual judgments to more theoretical geometric considerations can be seen in the progression through the identified group construction practices. Practices 1 through 4 are associated with a visual paradigm, whereas practices 5 through 9 are more appropriate to a theoretical one. The earlier conceptualizations are not entirely replaced by the more advanced mathematical ones. Rather, an increasing assemblage of ways of talking about a figure provide a deeper understanding. During their work on this topic, the team considers the geometric notion of perpendicularity in terms of a visual prototype of perpendicular line segments, the measurement of their angles of intersection, comparison to a standard model and the result of a specific construction sequence. They thus move through everyday conceptions, numeric considerations, symbolic representations and theoretical relationships—and associate them together.

Revision of the task for future use should emphasize the lessons learned from the analysis of the team's efforts on Topic 3. The challenge of constructing a perpendicular through a point other than the midpoint of a segment is too complicated to be given initially. It might make more sense to introduce this after the custom tool is created. Then one can ask for custom tools to be made for perpendiculars passing through points on the base line or even off the line. The instructions for testing the custom tool should have the students display a base line before using their tool to display a perpendicular to the base line. There should also be more explicit guidance about discussing the constraints defined by the construction of the perpendicular bisector.

Session 4: The Team Develops Tool-Usage Practices

Topic 4 is designed to consolidate the team's understanding of the construction approach by highlighting geometric relationships within triangles. It starts with a right triangle, in which two sides are perpendicular to each other. The first task is to define a custom tool that incorporates this relationship and automatically constructs right triangles. The second tab expands this approach to isosceles and equilateral triangles. The final tab challenges the students to assemble a hierarchy of types of triangles, based on their theoretical relationships among sides or angles. In our analysis of this session, we shall enumerate the adoption of *group tool-usage practices.*

Custom tools are construction tools that are defined by users of the GeoGebra application. GeoGebra provides an interface for a user to construct a figure and then encapsulate that construction in a new tool. For instance, there is no standard tool in GeoGebra for producing a right triangle. However, a user can define a custom tool for doing this and save the tool for future use in quickly generating right triangles. GeoGebra adds an icon to the menu bar for using the new custom tool. The VMT multi-user version of GeoGebra shares the new custom tools with all team members. By defining their own custom tools, students can gain insight into how GeoGebra tools are designed and how they maintain dependencies used in their definition.

Tab Right-Triangle

Team members start by saying (again) they do not know how to put a point on a line (step 1 in **Figure 20**; see lines 8, 9, 12 in **Log 27**), but then they soon construct a right triangle and a custom tool.

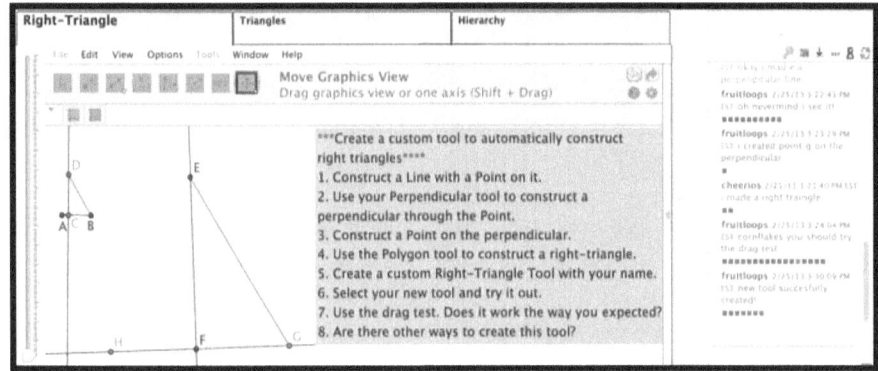

Figure 20. The Right-Triangle Tab.

Log 27. Constructing a right triangle.

Line	Post Time	User	Message
4	10:55.1	cheerios	hey
5	11:01.1	cornflakes	hi
6	11:55.9	fruitloops	hery
7	12:02.6	fruitloops	someone do step 1
8	14:02.1	cheerios	canu guys do it im not sure how to do it
9	14:17.5	fruitloops	i dont know how to
11	15:01.5	cornflakes	fruitloop udo it
12	15:23.2	fruitloops	i dont know how
13	17:46.3	cheerios	i did #1
14	18:07.1	cheerios	i made a line and put F on it
15	18:45.3	fruitloops	why is there 2 points on the line?
16	19:38.5	cornflakes	delete e
17	20:19.4	cheerios	is that good
18	22:34.6	fruitloops	so are you making the perpendicular line?
19	22:42.0	cornflakes	okay i mad e a perpendicular line
20	22:43.6	fruitloops	oh nevermind i see itt
21	23:29.5	fruitloops	i created point g on the perpendicular
22	23:40.1	cheerios	i made a right traingle
23	24:04.6	fruitloops	cornflakes you should try the drag test

Notice how quickly the team gets to work immediately after entering the chat room and announcing their presence by greeting each other. Fruitloops orients the team to step 1 of the tab's instructions and opens up the turn taking. Unfortunately, they each decline doing the first step, claiming they do not know how to "Construct a line with a point on it."

Eventually, Cheerios takes control and creates a line EF. While she says, "i made a line and put F on it" (line14), she actually made a line defined by points E and F. In the example figure, segment AB was first created and then point C was constructed to be on the segment. Because Cheerios created her line EF roughly parallel to the example line CD (which has no points on it other than its defining points C and D), it is likely that she was looking at CD rather than segment AB with point C as her model. Fruitloops wonders why there are two points on Cheerios' line EF, rather than just the one called for in the instructions. It is possible that Fruitloops said she could not do step 1 because she could not construct a line with just one point, but needed two points to define it with the line tool.

> *Group tool-usage practice #1: Use two points to define a line or segment.*

The instructions are ambiguous as to whether the one point is part of the line when it is created or is added onto a line that had already been created by two other points. The students' problem is not one involving lack of construction expertise, but rather difficulty in interpreting the instructions in relation to the example figure. They do not see the example figure as an instantiation of the construction procedure outlined in the tab instructions. If they did, they would figure out that one should first create a segment like AB and then construct a point like C on that segment. Instead, they try to copy the visual appearance of the example figure (e.g., mimicking line DE instead of segment AB). (Unfortunately, the instructions are misleading by calling for a line when a segment is illustrated.)

Cornflakes suggests to delete the extra point (line 16). Before her suggestion is posted, Cheerios has already deleted point F. This deletes the line that was partially defined by point F. So Cheerios creates a new line FG. G is located at a distance, so it is not visible on everyone's VMT screen. She then drags the isolated point E toward the line, as though she might try to put it on the line. She asks, "is that good" (line 17). In response, Fruitloops erases Cheerios' objects and constructs a new line EF. Then Cornflakes uses the GeoGebra perpendicular tool (not the group's custom tool from the previous topic) to construct a perpendicular to EF through point F. (The students were given access to the GeoGebra perpendicular tool in this topic because they had already learned how to create a perpendicular line using only the compass-and-straightedge tools, circle and segment in the previous topics.) Chat lines 18, 19 and 20 overlap in their typing. Fruitloops asks if Cornflakes is constructing a perpendicular in accordance with step 2 of the tab instructions. Then she sees it.

> *Group tool-usage practice #2: Use GeoGebra tools to construct perpendicular lines.*

Step 3 is Fruitloops' turn. She constructs a point G on the perpendicular line with no problem, and announces it in line 21. Cheerios takes her turn with step 4 and constructs triangle EFG with the polygon tool. In line 23, Fruitloops suggests that Cornflakes try the drag test on the triangle, taking her cue from step 7. Cheerios had released control, but now she takes it back. She moves the whole triangle back and forth a bit, without changing its size or proportions by dragging different vertices. It is not clear if she is just repositioning her triangle or responding to Fruitloops' suggestion to drag it.

Next, Fruitloops creates a custom tool for creating right triangles (Step 5). She defines its output as the right-triangle polygon EFG (line 26 in **Log 28**). The input is automatically defined as the two points of the base side EF. An icon appears on the tool bar for Fruitloops' custom tool. Cheerios then immediately does step 6 (line 28), using the custom tool to create four temporary right triangles (see the triangle built on segment EH in **Figure 21**). Cornflakes constructs four more triangles with the custom tool (built on segments IJ, KL, LK and JL).

> *Group tool-usage practice #3: Use custom tools to reproduce constructed figures.*

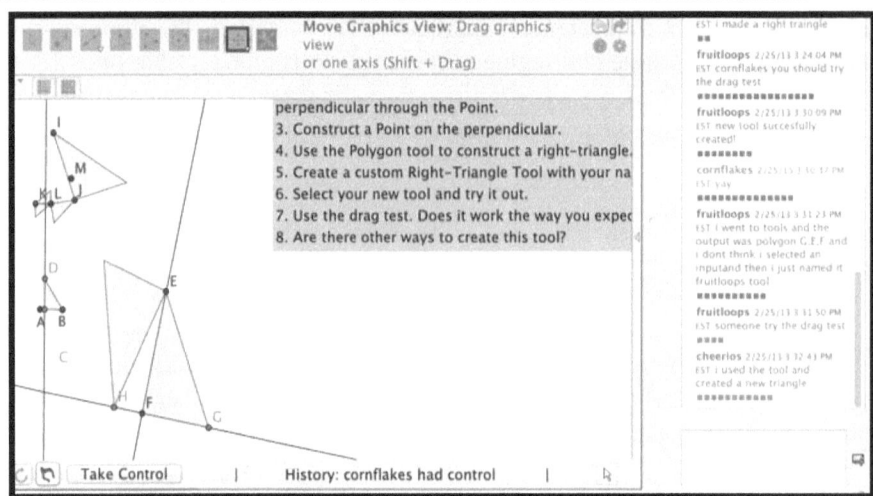

Figure 21. Some custom right triangles.

Log 28. Using the team's perpendicular tool.

24	30:09.5	fruitloops	new tool succesfully created!
25	30:37.3	cornflakes	yay
26	31:23.8	fruitloops	i went to tools and the output was polygon G,E,F and i dont think i selected an inputand then i just named it fruitloops tool
27	31:50.9	fruitloops	someone try the drag test
28	32:43.3	cheerios	i used the tool and created a new triangle
29	33:48.1	cheerios	cornflakes do the drag test
30	34:18.3	fruitloops	can i try dragging it?
31	34:25.1	cornflakes	sure
32	35:09.0	cheerios	can i try
33	35:33.0	fruitloops	sure
34	36:01.3	fruitloops	the lighter coloored points are restricted i think?
35	36:39.4	cornflakes	yes because they are already on the line
36	38:20.3	cheerios	are there any other ways to create this tool?
37	39:04.2	cornflakes	no i dont think so
38	39:11.2	fruitloops	i dont think so but i dont know for sure
39	39:38.8	cornflakes	are we ready to move on?

Fruitloops moves on to step 7, saying, "someone try the drag test" (line 27). Cheerios passes the task to Cornflakes (line 29). Then, everyone takes turns with the drag test. Cornflakes takes control and drags the vertices of triangle ABD, the original example triangle, not the ones that anyone created with the custom tool. Fruitloops wants to drag and compare the triangles. She drags the vertices of the three triangles that are now on the screen: her custom triangle EFG, Cheerios' custom triangle EHX and the example triangle ABD. Fruitloops drags them vigorously and systematically over a large range of sizes and orientations by dragging each vertex. Based on this dragging, Fruitloops concludes that "the lighter coloored points are restricted i think?" (line 34). Cornflakes explains this by pointing out that those points were constructed to be on an existing (perpendicular) line, so their movement under dragging is restricted to going back and forth on that line.

> *Group tool-usage practice #4: Use the drag test to check constructions for invariants resulting from custom tools.*

Meanwhile, Cheerios takes a turn at dragging the triangles around. She drags vigorously, but it is unclear what she is looking for. Dragging is not just a matter of moving whole figures around the workspace. Effective dragging is guided by a question or conjecture about the relationships within a figure: e.g., does the right angle remain a right angle when the vertices defining it are dragged? Cheerios does not display such directed inquiry in her dragging here.

Cheerios then raises the question of the tab's final step 8: "Are there other ways to create this tool." The team does not think there are other ways. In a by-now-typical hedge about what she knows, Fruitloops says she is not, however, sure about this (line 38). So they move on.

What they did not notice during their dragging was that all the angles of the right triangles created with the custom tool are fixed and identical to the original triangle that was used to define the tool. Had they noticed this, they might have explored different procedures to define custom tools with different angles or even with drag-able angles. (Some custom tools created in GeoGebra produce figures that are not fully dynamic. For instance, if a defining point of the base line is also used to define the line perpendicular to the base, then the resultant right triangles will not be dynamic. Using a third point—either on or off the base line—will define a custom tool producing dynamic right triangles.)

Tab Triangles

This tab (**Figure 22**) is originally empty except for instructions. Here, the instructions are not presented in numbered steps. Nevertheless, Fruitloops initiates work in the tab by requesting "someone try step 1" (line 45 in **Log 29**). The task is to construct triangles with different constraints. At first, Fruitloops asks, "how do you construct them" (line 47). Cornflakes proceeds without hesitation to select the generic polygon tool and create a triangle that looks roughly like a prototypical equilateral triangle. Then Cheerios places points and connects them with segments to form a nondescript triangle. At this point, Fruitloops joins in (line 48) and creates a polygon to connect three points, forming a triangle that looks isosceles. All three students have reverted to the visual approach.

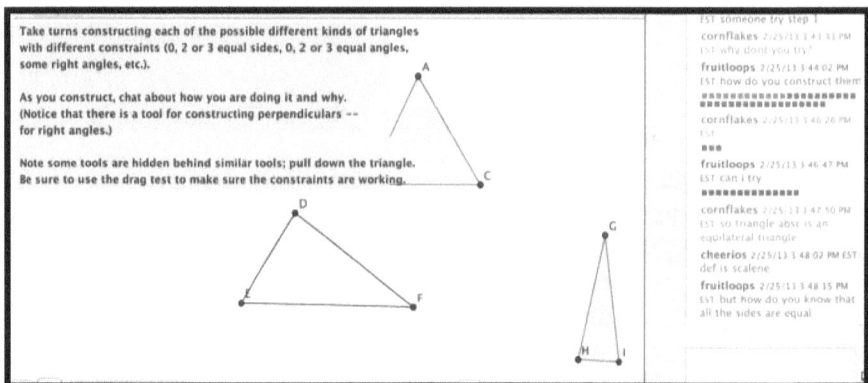

Figure 22. The Triangles Tab.

Log 29. Constructing different triangles.

45	40:39.2	fruitloops	someone try step 1
46	43:33.1	cornflakes	why dont you try?
47	44:02.8	fruitloops	how do you construct them
49	46:47.1	fruitloops	can i try
50	47:50.6	cornflakes	so triangle absc is an equilateral triangle
51	48:02.2	cheerios	def is scalene
52	48:15.8	fruitloops	but how do you know that all the sides are equal
53	48:34.7	cornflakes	it looks equilateral
54	49:02.7	fruitloops	but how can you prove it?
55	49:08.0	cheerios	with a ruler

56	49:32.2	fruitloops	did you measure it with a ruler?
57	49:36.5	cheerios	yes i did
58	49:45.3	cornflakes	sure
59	49:46.1	fruitloops	i dont believe you
60	49:52.7	cheerios	my finger is the ruler
61	50:11.6	fruitloops	but you finger isnt a proper measuring tool
62	50:22.0	fruitloops	NOT ACCURATE!
63	50:23.2	cornflakes	actually let me double check it
64	50:55.8	fruitloops	how do you double check?
65	50:58.3	cornflakes	they look pretty equal
66	51:43.1	cheerios	yea they do the only way is to measure it i guess
67	51:45.4	cornflakes	how do you tell?
68	52:20.3	fruitloops	but like when we did it with the circles and their radius we could prove it becuase of the equal radii but now i dont know how we can prove that its equilateral....

Cornflakes explains, "so triangle absc is an equilateral triangle" (line 50) and Cheerios adds, "def is scalene" (line 51).

However, Fruitloops raises the question of how Cornflakes can know that the three sides of her triangle are equal (line 52). This seems to be a potential move to a construction perspective, looking for geometric relationships. Cornflakes constructed her "equilateral" triangle with the generic polygon tool, the same way that Cheerios constructed her "scalene" triangle—that is, with no constraints. Cornflakes responds that she knows it because it looks that way (line 53). Then in response to Fruitloops' "but how can you prove it?" (line 54), Cheerios implies she measured the side lengths (line 55). Measurement is often an intermediate approach between purely visual and theoretical.

There follows a beautiful exchange defining the team's transitional status between visual appearance and proof by dependencies. Cheerios and Cornflakes rely on the appearance and measurement. Fruitloops points to the lack of precision in measuring and wants to know how to prove it in the sense of Euclid's proof of the equilateral-triangle construction with equal radii. Her original hesitation in line 47 about how to construct the various triangles may have envisioned a construction process like that of the equilateral triangle in their second session, rather than the simple use of the polygon tool or segment tool that Cornflakes and Cheerios used.

The team engages in a playful interchange about how they measured with their fingers and how this was unacceptably inaccurate. Cornflakes reiterates her reliance on how things look equal and Cheerios sticks to measurement as the only way to know.

In line 68, Fruitloops recalls the constructions of the equilateral triangle and the perpendicular bisector in Topic 2, using circles and their equal radii to prove equality of segment lengths. She says, "but like when we did it with the circles and their radius we could prove it becuase of the equal radii but now i dont know how we can prove that its equilateral." For her, special construction made proof possible. Simply drawing a triangle using the generic polygon tool relied on rough appearances and imprecise manual adjustments, with no basis for proving equality. As in the previous topic, Fruitloops tries to solve the current problem by applying their approach on a related past problem, specifically the equilateral-triangle construction using circles.

Cheerios proposes overlaying a grid of equally spaced lines on top of the triangles (line 70). GeoGebra allows a user to display a grid across the workspace. Since each cell of the grid is of equal length and width (line 74 in **Log 30**), it could be used to precisely measure triangle sides, especially if the endpoints are snapped to align to the grid.

Log 30. Measuring an equilateral triangle.

69	52:37.5	cheerios	i think we need a line
70	52:45.1	cheerios	or a grid
71	53:09.7	fruitloops	how would a crid really help?
72	53:31.6	fruitloops	grid*
73	53:48.6	cornflakes	I am not sure how make sure mathematically that triangle abc is equilateral
74	54:44.8	cheerios	each box is equal length and width
75	54:59.0	cornflakes	right so a grid would help
76	55:04.4	fruitloops	you can try it but i dont know for sure
77	55:08.4	cheerios	i think it would
78	55:29.9	fruitloops	try it then
79	56:03.3	cheerios	does everyone have a grid
80	56:05.4	fruitloops	i think we all have a grid now right?
81	56:40.7	cornflakes	correct6]
82	58:15.0	fruitloops	what do we do now?
83	58:39.4	cornflakes	im not sure

84	59:17.2	fruitloops	i dont know wbhat tio do i just moved it around
85	00:05.2	cheerios	you have to line it up so point a on an intersection and then see how far away point b and c are from the line a is on
86	00:19.7	fruitloops	yeah try to show it
87	00:26.9	cornflakes	can you show it?
88	00:36.6	cheerios	i just did
89	00:45.8	fruitloops	where?
90	01:04.9	cornflakes	are we ready to move on?
91	01:16.5	cheerios	a is on the intersection and pint b and c is one box away
92	01:31.9	fruitloops	i kind of understand
93	01:48.0	fruitloops	we can try hierarchy if you want
95	01:54.6	cornflakes	ok i get it
97	01:57.1	cheerios	okay

The team decides to try using a grid. They each turn on the grid display (line 80). Cornflakes drags the vertices of triangle ABC around to line it up with the grid. She does not seem convinced that this worked because in response to Fruitloops' question of what to do now that they have the grid, Cornflakes says, "im not sure" (line 83).

Fruitloops then drags the vertices of triangle ABC strongly and sees that it can be distorted from an equilateral triangle down to a flat line. However, she does not comment on the fact that triangle ABC does not always look equilateral. The team is using the dragging not to vary the figure to see what remains fixed (the drag test of the theoretical approach of dynamic geometry)— rather, they drag to get the figure to look equilateral (the visual approach of drawing). This may be a sign of a misconception within the team's understanding of dynamic geometry. Whereas dynamic geometry is concerned with geometric relationships that remain *invariant* when points are moved dynamically, the students may be concerned with geometric relationships that are *possible* when points are moved. They are seeing that the polygons they constructed *can* be isosceles or equilateral through dragging, not that they *must* be regardless of dragging. Such a misconception may form an intermediate stage, on their way to a mathematically accepted conception that is difficult to understand immediately from a non-dynamic or visual perspective.

Fruitloops is now in a quandary (line 84). Cheerios says they have to line the points up on the grid (line 85) in order to show that they can be located at equal distances from each other—and so she arrays them on the equally spaced intersections of a grid (line 88). As can be seen in **Figure 23**, side BC is two grid units long, but sides AB and AC are longer—they are the diagonals of a right triangle that is one unit by two units. Without coming to a clear consensus, the team moves on.

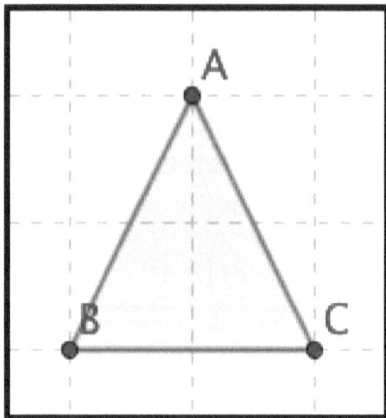

Figure 23. The grid.

Tab Hierarchy

The Hierarchy Tab (**Figure 24**) is intended to spark discussion about how an equilateral triangle is a special case of an isosceles triangle, which is a special case of a scalene triangle. A scalene triangle can be dragged to look like an isosceles or equilateral triangle, but it can also be dragged to look different. In contrast, a triangle that is constructed to be equilateral will always look equilateral no matter how its vertices are dragged.

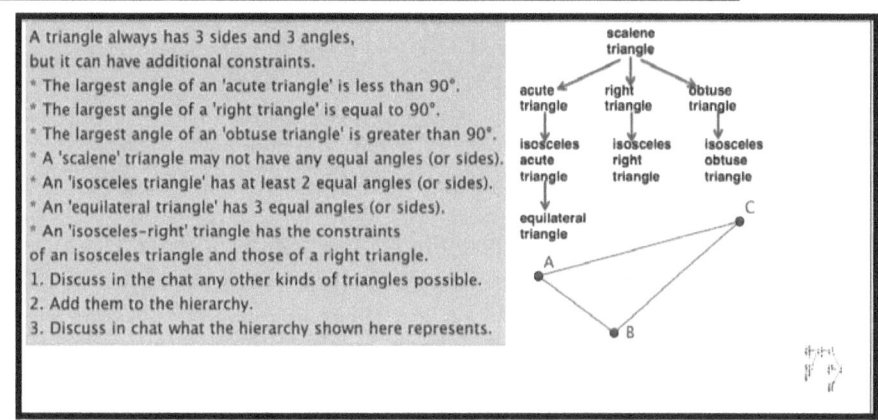

Figure 24. The Hierarchy Tab.

The team spent about nine minutes unproductively in this tab. They discovered a tiny image of the hierarchy graphic (left in the tab by mistake) and they dragged it around the workspace. They also dragged the textbox with the instructions around and created a couple of new points.

The team only made one chat posting for this tab (line 100 in **Log 31**). Fruitloops suggested ignoring the tiny image and discussing the hierarchy chart of different kinds of triangles. After another two minutes of silence, the team ended their session.

Log 31. The only posting in the tab.

100	09:03.4	fruitloops	okay so lets only look at the lager chart

The team was clearly not ready to engage in the discussion intended for this tab. In previous tabs, they had not recognized that a dynamic triangle could sometimes look equilateral, sometimes isosceles, sometimes right, sometimes obtuse and other times simply scalene. They still seem to associate these categories with static appearances, rather than with dynamic relationships or dependencies. Student conceptions are neither logical nor consistent during the learning process. They cannot reliably be predicted by curriculum designers without testing activities with student groups in realistic situations and closely analyzing the reactions.

Summary of Learning in Session 4

In their fourth session, the team's discourse is quite variable; in some tabs they discuss what they are doing and reflect on what happens—generally following the prompts in the instructions. At other times, they do not know what to do in response to a prompt or an instruction and they remain silent. Their understanding of construction in dynamic geometry is also variable. It varies from tab to tab, from moment to moment and from student to student. This reflects the uneven, episodic evolution of development, as the team starts to move from appearances to dependencies in lurches forward and back. Sometimes they seem to have problems with tasks they have accomplished in the past and at other times, they seem to do new things without difficulty.

Through increased practice with constructing figures in GeoGebra, the team becomes generally more skilled at using the tools—both built-in GeoGebra tools like the perpendicular tool and custom tools that the team itself defines. We have identified some group tool-usage practices that the team adopted during this session, refining their use of the GeoGebra tools:

1. *Use two points to define a line or segment.*

2. *Use special GeoGebra tools to construct perpendicular lines.*

3. *Use custom tools to reproduce constructed figures.*

4. *Use the drag test to check constructions for invariants resulting from custom tools.*

These practices contribute incrementally to the team's development of mathematical group cognition. The ability to use the tools of dynamic geometry is not acquired automatically. It requires practice in a variety of constructions. There are multiple ways to accomplish the same thing (like defining a line with existing or new points), and people or groups must learn when the different options are appropriate. One must know how to select the right tool. GeoGebra opens up the tool selection process by supporting the definition of new custom tools. The use of custom tools, in particular, calls for confirming expected invariants with the drag test.

At the end of this session of the Cereal Team, the team improves in some ways:

i. Its *collaboration* is generally quite good. They get started quickly and they proceed systematically through the steps of the instructions.

ii. Its productive mathematical *discourse* is inconsistent. They still often talk about visual appearances rather than mathematical relationships. However, they do engage in a good discussion, which confronts and compares these two approaches, during this session.

iii.   The team learns to define and use a *custom tool*. However, it is not clear that this provides them insight into how GeoGebra tools are designed or how they could define their own tool differently using alternative constructions.

The team's ability to *identify and construct dynamic-geometric dependencies* is, consequently, also inconsistent. The team becomes much more acquainted with the drag test in this session, although they are not always clear about what it shows or could be used for:

a.   *Dynamic dragging* is undertaken a number of times. Sometimes the students engage in quite vigorous dragging. However, they have not adopted it as a regular check on their construction validity.

b.   *Dynamic construction* is emphasized with the first tab of this session. The team manages to define custom tools—which institutionalize construction processes and their associated dependencies—with no problem.

c.   *Dynamic dependency* in GeoGebra is still unclear. However, Fruitloops seems to be getting the idea.

We will observe how each of these progresses further in the remainder of the sessions.

The team's work on each of the tabs suggests potential revisions of the curriculum. For the Right-Triangle Tab, there might be a way to motivate discussing alternative constructions, with their pros and cons. This might be too advanced at this point for a group like the Cereal Team. However, such considerations are important for design-oriented thinking. For the Triangles Tab, some guidance on using construction techniques such as circles for imposing constraints may be needed. It is particularly important that students understand how a circle can be used to constrain the side lengths of an isosceles triangle. The Hierarchy Tab could stimulate interesting discussions about geometric relationships, but only once students understand the dynamic character of figures, such that whether or not a triangle that appears to be isosceles can be dragged into a scalene triangle depends on constraints established during its construction.

Session 5: The Team Identifies Dependencies

In this section, we review the work of the group on Topic 5, a problem of inscribed equilateral triangles, squares and hexagons. We analyze the chat log of the team working on this tab in even more detail than the other sessions—pointing out a series of *group dependency-related practices*. The Cereal Team spends two hour-long sessions on Topic 5. In the first session, they work on the inscribed triangles and in the second session (see next chapter) on the inscribed squares. Although they look at the inscribed hexagons tab, they never have time to work on it. The three tabs of Topic 5 are shown in **Figure 25**.

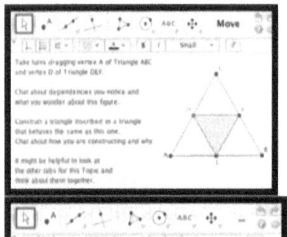
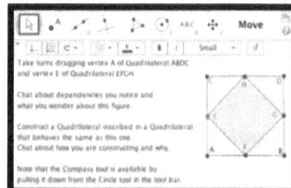

Topic 5 is explicitly about dependencies. It prompts the students to explore and discuss the dependencies in figures of inscribed polygons, especially equilateral triangles and squares. Then it challenges them to construct their own figure with the same behavior—e.g., remaining equilateral and inscribed when vertices are dragged. The solution involves the use of GeoGebra's compass tool to construct segments whose lengths are dependent on each other so that geometric relationships within the figure are invariant when dragged.

The Inscribed Triangles Tab

Starting with **Log 32**, all of the chat postings of the three students for this dual session are listed in this book. The team begins by following the instructions in the opening tab: "Take turns dragging vertex A of Triangle ABC and vertex D of Triangle DEF."

Log 32. The team explores the triangles.

Line	Post Time	User	Message
3	11:53.8	fruitloops	heyyyyyyyyyyyyyyy
4	13:06.0	cornflakes	hi
5	13:30.9	cornflakes	i will go first
7	18:09.6	fruitloops	when i move vertex a the whole triangle of abc moves
8	18:43.8	cornflakes	when i moved point c the triangle stayed the same and either increased or decreased in size, butit was equivalent to the original triangle
9	18:52.8	fruitloops	but when i tryed to move vertex d, it couldnt go behond triangle abc
10	18:54.4	cheerios	does the inner triangle change its shape when u move vertex a
11	19:34.3	fruitloops	try moving it...
12	20:38.9	cheerios	nvm it doesnt
13	22:43.7	fruitloops	yeah when you move vertex a, the inner triangles changes size but never shape
14	22:54.3	cornflakes	yes
15	23:35.2	fruitloops	can i try to make the circle equilateral triangle fist?
16	23:38.2	cheerios	yes
17	23:53.4	cornflakes	sure
18	24:11.5	fruitloops	wait, fist we should talk about the other vertexes
19	24:23.7	cornflakes	yes
20	24:28.8	cheerios	agreed
21	24:48.8	fruitloops	so cheerios since you have control what happens when you move the different vertexes?
22	25:27.0	cheerios	when you move vertex a triangle dfe dont move at all it just becomes smaller when you shrinnk the big triangle and vice versa
23	25:43.5	cornflakes	
24	25:44.2	cornflakes	
25	25:56.5	fruitloops	what about point e? c? F?

The students drag points A and D. They quickly see that the interior triangle is confined to stay inside triangle ABC and that both triangles retain their shape when dragged. Fruitloops is eager to start constructing an equilateral triangle using circles. They had learned the construction in Topic 2 and been reminded of it in class earlier that day. However, Fruitloops reconsiders and suggests that they explore further by dragging the other points. This proposes a first group practice for exploring dependencies, building on practices previously adopted by the team for tool usage and dragging.

> Group dependency-related practice #1: Drag the vertices of a figure to explore its invariants and their dependencies.

In **Log 33**, the students start to discuss the dependencies in more detail. They note that points C, E and F are "sconstrained or restricted," so these points are not free to be dragged. They also note that dragging point D will move points E and F. This will turn out to be a key dependency, although the students do not discuss it as such.

Log 33. The team discusses dependencies

26	26:41.0	cornflakes	ecf arent moving
27	27:00.7	fruitloops	point c e and f cant move
28	27:52.6	cornflakes	because they are sconstrained or restricted
29	27:53.4	fruitloops	point d can only make point f and g move but nothing else
30	28:29.3	cornflakes	yea
31	28:50.5	fruitloops	okay want to try to conssrtuct it?
32	29:02.0	cheerios	yup
33	29:07.3	cornflakes	sure

The team is now ready to begin the construction task. Fruitloops begins the construction with a segment GH and two circles of radius GH centered on points G and H, respectively. Fruitloops gets stuck at line 34 and Cheerios takes over, drawing the triangle connecting point I (at the intersection of the circles) with points G and H. They thereby adopt the procedure from the earlier topic of constructing an equilateral triangle as a general practice for establishing dependencies.

> *Group dependency-related practice*
> *#2: Construct an equilateral triangle*
> *with two sides having lengths*
> *dependent on the length of the base,*
> *by using circles to define the*
> *dependency.*

Fruitloops wants to remove the circles, but seems to understand in line 36 of **Log 34** that they cannot erase the circles without destroying the equilateral triangle. Cornflakes hides the circles by changing their properties—which maintains the dependencies that are defined by the circles, while clearing the circles from view.

> *Group dependency-related practice*
> *#3: Circles that define dependencies*
> *can be hidden from view, but not*
> *deleted, and still maintain the*
> *dependencies.*

Log 34. The team constructs the first triangle.

34	30:27.0	fruitloops	what should i do next?
35	32:22.9	fruitloops	so how do we get rid of the circles then?
36	32:54.4	fruitloops	if we cant delete them, what do we do?
37	34:37.3	fruitloops	so i think triangle igh is like triangle abc
38	36:30.3	fruitloops	now that the first triangle is good, what should we do?

In line 38, Fruitloops suggests that they have succeeded in replicating the outer triangle. Then in **Log 35**, Fruitloops makes explicit that their previous observation about movement of point D affecting points E and F implies a dependency that may be relevant to their construction task. Cheerios and Cornflakes express interest in this line of argument. They all agree to proceed with trying constructions in order to figure out just what needs to be done. As with designing the exterior triangle, the results of dragging provide an impetus for construction, but not a blueprint. The participants launch into a trial-and-error process, guided by some vague ideas of things to try.

Log 35. The team experiments.

39	47:48.2	fruitloops	d moves but f and e dont

40	48:04.2	fruitloops	so both f and e are dependent on d
41	48:15.9	cornflakes	right
42	48:18.3	cheerios	so what does that mean
43	48:37.5	fruitloops	so if we make a line and use the circle thing, maybe we can make it somehow
44	49:09.8	cheerios	lets try
45	49:14.4	fruitloops	how?
46	49:29.8	cheerios	and we will jsut figure it out .. by making the line thing
47	50:18.4	cheerios	f and e are restricted
48	51:20.0	fruitloops	we can make their d point by just using a point tool on our triangle to make point j
67	11:35.4	fruitloops	so what ere you dong now?

They begin their trial with the knowledge that point D is freer than points E and F, which are dependent on D. Therefore, they decide to start by constructing their equivalent of point D on a side of their exterior triangle.

> *Group dependency-related practice #4: Construct a point confined to a segment by creating a point on the segment.*

Note the gap of about 20 minutes from line 48 to the next chat posting. This was a period of intense experimentation by the three students. Unfortunately, they did not chat about what they were doing during this period. We have to look at a more detailed log and step through the VMT Replayer slowly to observe what they were doing.

The logs shown so far in this book have all been filtered to show only text-chat postings. **Log 36** is taken from a more detailed view of the log spreadsheet including GeoGebra actions, such as selecting a new GeoGebra tool from the tool bar or using the selected tool to create or change a GeoGebra object. It also includes system messages, such as announcing that a user has changed to view a different tab. (The GeoGebra actions are not assigned line numbers. The system messages are assigned line numbers; they account for the line numbers missing in the other chat logs in this section.)

Log 36. The team views other tabs.

	15:57:10	Geogebra: Triangles	cheerios	tool changed to Move
	15:57:27	Geogebra: Triangles	cheerios	updated Point A
	15:57:28	Geogebra: Triangles	cheerios	tool changed to Move Graphics View
	15:58:35	Geogebra: Triangles	cornflakes	tool changed to Move
	15:58:39	Geogebra: Triangles	cornflakes	updated group of objects G,H
	15:58:42	Geogebra: Triangles	cornflakes	tool changed to Move Graphics View
47	15:59:08	system	cornflakes	Now viewing tab Squares
48	15:59:13	system	cornflakes	Now viewing tab Triangles
49	15:59:17	system	cornflakes	Now viewing tab Hexagons
50	15:59:21	system	fruitloops	Now viewing tab Squares

In this excerpt from the detailed log, we can see that Cornflakes uses certain GeoGebra tools to change (drag) specific objects in the construction. We also see that Cornflakes—like Fruitloops and Cheerios—looks at the other tabs. This is just a brief sample of what took place during the 20 minutes. There were actually 170 lines in the detailed log for that period, listing all the actions taken with GeoGebra tools (selecting a tool, creating a new object, dragging an existing object, etc.). During all this activity, the students make very little obvious progress on their construction. They construct some lines, circles and points. They engage in considerable dragging: of the original figure, of their new triangle and of their experimental objects. They also each look at the other tabs.

In particular, the students played with several circles, experimenting with how to define them with two points (either already existing or newly created as part of the circle construction), how to drag them and how to drag existing points onto them. Because mathematical cognition is mediated by tool use, the students had to better understand how the circle tool and the related compass tool worked before they could understand the problem of inscribed triangles and discuss an approach to constructing inscribed triangles. The key to analyzing the dynamic figure of the triangles involves dependencies among equal lengths as defined by radii of circles. So the students have to know how circles and their radii are defined and how they work to establish length dependencies. Apparently, their 20 minutes of exploration of circle and compass construction allowed the students to better understand the inscribed-triangles construction.

Finally, Cheerios provides the key analysis of the dependency: AD=BE=CF (lines 68 to 74). The others immediately and simultaneously agree with this analysis. In **Log 37**, Cheerios goes on to project this dependency onto their construction in line 75. (This chat excerpt is also shown in **Figure 26** in the context of the state of the figure after the team finished constructing triangle KMR inscribed in triangle GHI.)

Log 37. The team makes a key observation.

68	18:30.0	cheerios	as i was movign d segment da is the same distance as segment be
69	18:52.0	cheerios	and also cf
70	19:41.6	cheerios	our kg is the same as ad
71	20:06.3	cornflakes	agrreeed
72	20:06.5	fruitloops	i agree
73	21:21.8	cheerios	there should be a point on segment gh which is the same distance as kg and also between segment uh
74	22:00.5	cheerios	it should be ih not uh
75	23:39.9	cheerios	so i used the compass tool and measured kg and used point i as the center and created a circle

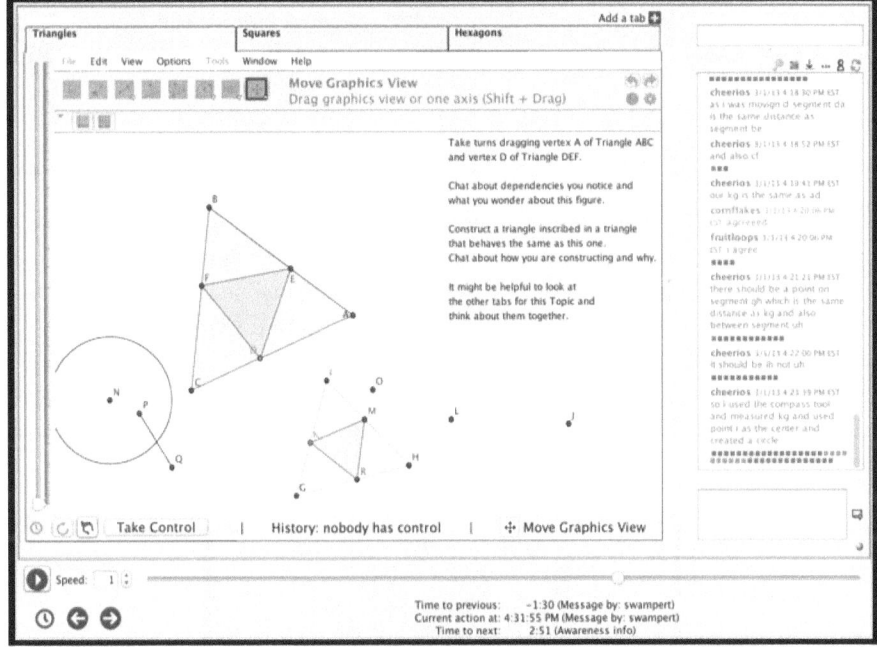

Figure 26. The construction after Fruitloops finished.

What was particularly striking in the team's successful construction of the inscribed triangles was that on first appearance it seemed like the team's insightful and skilled work was actually done primarily by the student who until then had seemed the least insightful and skilled. If one just looks at the chat postings, Cheerios does all the talking and Fruitloops (who is usually the most reflective and insightful) and Cornflakes (who explores the technology and often shows the others how to create geometric objects) simply register passive agreement. However, the actual GeoGebra construction actions tell a far more nuanced story.

Cheerios observes Fruitloops experimenting with the use of the GeoGebra compass tool just before Cheerios takes control and makes her discovery that AD=BE=CF. Cheerios continues to manipulate Fruitloops' construction, involving a circle whose radius was constructed with the compass tool to be dependent on the length of a line segment. Then Cheerios very carefully drags points on the original inscribed-triangle figure to discover how segments BE and CF are dependent upon the length of segment AD, refining prior movements by the other students. The dynamic relationship between the side lengths becomes visually salient as she increases the size of the triangles or their orientation and as she drags point D along side AC.

> *Group dependency-related practice*
> *#5: Construct dependencies by*
> *identifying relationships among*
> *objects, such as segments that must be*
> *the same length.*

Cheerios has a sense that the compass tool should be used to measure segment KG, but she does not quite understand how to make use of that tool. The students had earlier in the day watched a video of such a construction— using the compass tool to copy a length from one line onto another line—in class in preparation for this topic, and had previously been introduced to it in Topic 2. Cheerios uses the compass tool to construct a circle around vertex I with a radius equal to the length GK (see **Figure 27**). However, she does not place a point where that circle intersects side HI, to mark an equal length.

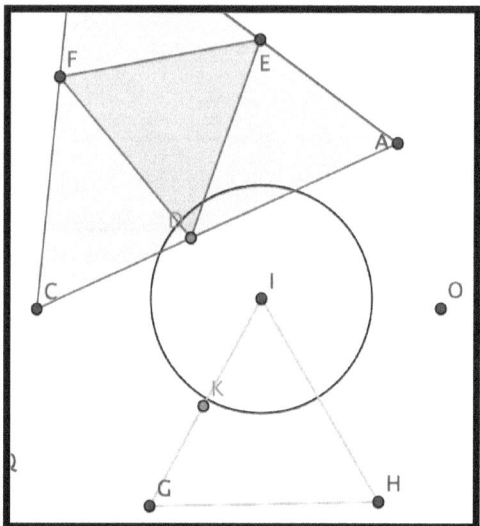

Figure 27. The construction after Cheerios finished.

Next, Cornflakes steps in to help Cheerios carry out the plan. Cornflakes takes control of the construction, places a point, M, where Cheerios' circle intersects side HI and then repeats the process with the compass to construct another point, R, on the third side of the exterior triangle (see **Figure 28**).

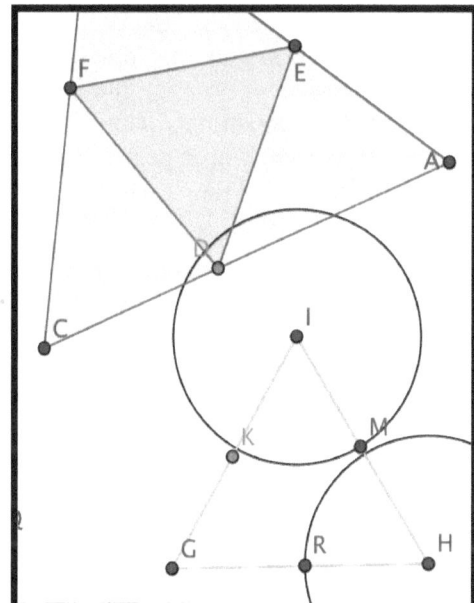

Figure 28. The construction after Cornflakes finished.

> *Group dependency-related practice*
> *#6: Construct an inscribed triangle*
> *using the compass tool to make*
> *distances to the three vertices*
> *dependent on each other.*

Fruitloops then takes control and uses the polygon tool to construct a shaded interior triangle, KMR, connecting Cornflakes' three points on the sides of the exterior triangle (see **Figure 26** above). She then conducts a drag test, dragging points on each of the new triangles to confirm that they remain equilateral and inscribed dynamically, just like the example figure. At that point, the students have been working in the room for over an hour and end their session, having succeeded as a team.

> *Group dependency-related practice*
> *#7: Use the drag test to check*
> *constructions for invariants.*

Thus, all three group members not only verbally (in chat) agreed with the plan, but they also all participated in the construction (in GeoGebra). The team as a collaborative unit thereby accomplished the solution of the problem in tab A. It is surprising that Cheerios noticed the key dependency among the dynamic locations of the vertices of the inscribed triangle, because she seemed to be the last to recognize such relationships in previous topics. She also articulated the implications in terms of a plan for constructing the interior triangle in a way that would maintain the dynamic constraints. Cornflakes and Fruitloops not only agreed with the plan through their chat postings, but also displayed their understanding by being able fluidly and flawlessly to help Cheerios implement successive steps of the plan.

The surprising thing is that Cheerios was the team member who made the conceptual breakthrough in noticing that AD=BE=CF. Cheerios was particularly quiet through most of this session. She did not contribute anything significant prior to this breakthrough. Apparently, she was, however, paying careful attention and now understanding what the others were saying and doing. Despite long periods of chat silence, there was some tight collaboration taking place, for instance in **Log 35**. At line 39-40, Fruitloops says, "d moves but f and e dont. so both f and e are dependent on d." Cheerios responds to this with, "so what does that mean" (line 42). This may orient Cheerios to focus herself on how points E and F are dependent upon (i.e., move in response to movements of) point D. Two minutes later, after encouraging the group to do explorative constructions, Cheerios repeats "f and e are restricted" (line 47). Then, when she is dragging the inner triangle around, she sees how F and E follow the movements of D, maintaining the equality of their corresponding segments.

> *Group dependency-related practice #8: Discuss relationships among a figure's objects to identify the need for construction of dependencies.*

Through the adoption of a sequence of dependency-related practices as joint group practices, skills previously associated with one team member become shared with the rest of the team. Other group practices—such as collaboration group practices of joint attention, co-presence and intersubjectivity—also contribute to such sharing. Behaviors that once seemed to be associated with the individual perspective of one student are now occurring in the actions of the others and of the joint group activity. In these ways, contributions of individuals are mediated by resources from the curriculum and by the interaction of the group to result in both collaborative learning and individual development. In the details of the team's interaction, we can observe the interpenetration and mutual influence of developmental processes at the individual (personal), group (team) and community (mathematical) levels.

Summary of Learning in Session 5

In this session, the team makes a dramatic leap forward in displaying an orientation to and understanding of dynamic construction incorporating dependencies. We have identified eight group dependency-related practices that the team adopted during this session:

1. *Drag the vertices of a figure to explore its invariants and their dependencies.*

2. *Construct an equilateral triangle with two sides having lengths dependent on the length of the base, by using circles to define the dependency.*

3. *Circles that define dependencies can be hidden from view, but not deleted, and still maintain the dependencies.*

4. *Construct a point confined to a segment by creating a point on the segment.*

5. *Construct dependencies by identifying relationships among objects, such as segments that must be the same length.*

6. *Construct an inscribed triangle using the compass tool to make distances to the three vertices dependent on each other.*

7. *Use the drag test to check constructions for invariants.*

8. *Discuss relationships among a figure's objects to identify the need for construction of dependencies.*

These are important practices for constructing figures with dependencies in GeoGebra. While the team has previously used some of them tentatively, they use them quite effectively and confidently in this session, with the exception of the twenty-minute period of intense, but chat-free exploration of GeoGebra construction techniques, which led to their being subsequently able to do the construction.

At the end of the fifth session of the Cereal Team, we see how the team improves:

i. Its *collaboration*, in that the team members get to work together quickly and effectively, discuss core issues of the task and fluidly complete each other's construction steps. Their individual skills merge into collaborative group practices.

ii. Its productive mathematical *discourse*, in pointing out a key dependency in the given figure, which is central to completing the mathematical task.

iii. Its ability to use dragging to identify dynamic-geometric dependencies—particularly that AD=BE=CF—and to use the compass *tool* to construct the same dependencies.

We also track the team's fluency with *identifying and constructing dynamic-geometric dependencies*:

a. *Dynamic dragging* becomes a major activity in exploring the given figure and the students' own attempts at construction.

b. *Dynamic construction* is the central activity in this session. The breakthrough is to come up with an insightful and effective plan in advance of construction. This was facilitated by the students' increased familiarity with the GeoGebra circle and compass tools.

c. *Dynamic dependency* in GeoGebra finally comes to the fore, as Cheerios states the need to maintain the dependency of AD=BE=CF through a construction using the compass tool.

The inscribed-triangles task has been used successfully many times before in the VMT Project. It naturally combines dragging to explore an example figure and then the use of the compass tool to construct discovered dependencies (Öner, 2008). It can serve usefully for formative assessment of a team's understanding of dynamic geometry and its creative use of GeoGebra. It is rare for an individual to solve the challenge in less than an hour, but well functioning teams—like the Cereal group—can often do it collaboratively.

Session 6: The Team Constructs Dependencies

Three days later, the team reassembles for Session 6 in the same chat room to continue work on Topic 5. They had hurriedly completed the construction of the inscribed triangles, but had not had a chance to fully discuss their accomplishment. Furthermore, they had not had any time to work on the other tabs. The class was given another one-hour session to work on Topic 5. During class earlier that day, the teacher had asked the Cereal Team of students to present their solution of the inscribed-triangles problem to the rest of the class on a large smart-board projector, since they were the only group in the class to successfully complete this construction during Session 5. In this session, we identify additional *group dependency-related practices*.

The Inscribed Triangles Tab, Continued

Cheerios resumes the team's discussion by announcing that they have to explain what they did. This directive had come from the teacher before the session started.

Back in the VMT chat room, Cheerios begins to explain (lines 92, 93 and 96) what they had done at the end of the previous session. Cornflakes joins in. Fruitloops decides not to post her comments (lines 96, 97) When Cornflakes says (in line 98 in **Log 38**), "you had to make the point between the two circles," Cheerios clarifies (line 100): "not between the circles (but) where the segment intersect(s) with the circle." Cornflakes may have been confusing the construction of the first triangle (with intersecting circles) with that of the interior triangle (with the compass circle intersecting the triangle side). At any rate, Cheerios uses the formal mathematical terms, "segment" and "intersect," and Cornflakes indicates that they are in agreement on what took place in the construction.

Log 38. The team explains its construction.

92	16:18.2	cheerios	we have to explain what we did
93	19:49.0	cheerios	so first u have to plot a random point on the triangle we used k . then i realised the distance from kg is the same as im and rh

94	20:41.4	cornflakes	right
95	22:51.2	cheerios	then you have to use the compass tool in are case are the length of are radius is kg so then we clicked those 2 points and used vertex i as the center the way to plot are second point of are triangle is where the circle and segment ih intersect
96	22:52.2	fruitloops	
97	22:52.5	fruitloops	
98	23:48.8	cornflakes	yes you had to make the point between the circles
99	23:53.2	cheerios	and then we repeated that step with the other side and h was the center
100	24:21.0	cheerios	not between the circles where the segment intersect with the circle
101	25:27.5	cornflakes	yea same thing

Fruitloops then raises a question about the dependencies among the points forming the vertices of the interior triangle (line 105 of **Log 39**). She notes that the two points constructed with the compass tool are colored black (or shaded dark), an indication of dependent points.

> *Group dependency-related practice #9: Points in GeoGebra are colored differently if they are free, restricted or dependent.*

Log 39. The team explains the dependencies.

102	25:52.4	cheerios	its differnt
103	25:58.5	cheerios	different*
104	26:00.0	cornflakes	yes i know
105	26:04.4	fruitloops	so then why are point m and r shaded dark and don tact the same as k
106	26:14.6	cheerios	they are restricted
107	26:35.6	fruitloops	but whyy???????
108	26:45.8	cornflakes	yeah if its a darker its restricted i think
109	26:52.1	cheerios	yes
110	26:56.4	cheerios	correct

111	28:31.0	fruitloops	but why are m and r restricted but k isnt?
112	30:33.3	cornflakes	because the invisible cirlcels are still there
113	31:35.0	fruitloops	okay so its because we made k by just using the point tool and putting it on the line but with m and r we maade it through using circles so technicaly, the circle is still there but its hidden but we just dont see it.
114	31:44.5	cornflakes	right
115	31:49.5	cheerios	yup
116	31:57.0	fruitloops	and i think we can move on because i understand it well. do you guys get it?
117	32:04.9	cheerios	yes
118	32:10.9	cornflakes	sure

The students all agree that the two points m and r are different from the first one, k, in terms of being more restricted. However, Fruitloops requests more of an explanation about why this is. Cornflakes explains that the circles are still in effect, constraining the locations of the points they help to define even though someone had hidden the circles formed by the compass tool by changing the properties of the circles to not show themselves. Fruitloops then explicates that the difference is that the first point was just placed on a side of the larger triangle. (So, it can be dragged, as long as it stays on the side.) However, the more completely restricted two points were constructed with the circles. (So, they must stay at the intersections of the circles with their sides; they cannot be directly dragged at all). Although the compass circles have been hidden from view, the dependencies that they helped to define (the intersections) are still in effect. One could go on to discuss how moving the first point will alter the lengths that define those circles and therefore will move the other points, but the students state that they all understand the reason why the different points are colored differently and have different dependencies. They are ready to move on and they all change to the tab with inscribed squares.

This confirms that the team has understood how the use of the compass tool can impose relationships of dependency (such as to maintain that "the distance from kg is the same as im and rh" as Cheerios says in line 93), even if the compass' circles are invisible. This is an important insight into the design of dependencies in dynamic geometry and the corresponding use of GeoGebra tools.

> *Group dependency-related practice #10: Indications of dependency imply the existence of constructions (such as regular circles or compass circles) that maintain the dependencies, even if the construction objects are hidden.*

We see here how the team's conceptual understanding of dependency, their practical understanding of how to create relationships of dependency, their understanding of how to interpret attributes of displayed figures in GeoGebra and their detailed pragmatic understanding of how to use specific GeoGebra tools are tightly interconnected. This is what we mean by saying that the group's ability to make sense of problems and strategize about how to approach their solution is "mediated" by the meaning they make of the available tools for action. The meaning of dependency for the team is constructed and expressed in terms of their enactment of the tools, like the compass tool.

The Inscribed Squares Tab

In the new tab, the students start again by dragging to explore dependencies. In **Log 40**, Fruitloops does the dragging and reports three classifications of points.

Log 40. The team explores the square.

122	32:52.2	fruitloops	can i try dragging it?
123	32:56.1	cheerios	yea
124	33:44.3	cheerios	u can try now fruitloops
125	35:03.9	fruitloops	so b and a move and points c,h,d,g, and f dont move
126	35:28.2	fruitloops	and e is restrricted
127	35:34.3	cornflakes	E IS RESTRICTED
128	35:58.6	fruitloops	do how do we create a square like the outer square?
129	36:54.4	cheerios	we have to talk about the dependencies and stuff
130	37:01.3	cheerios	read the instructions

Two points of the outer square can move (freely), one point of the inner square is "restricted" (constrained) and the other points don't move (are dependent). Cornflakes echoes the "E IS RESTRICTED" as though to elicit discussion of this special status (line 127). Cheerios' chat posts also try to insist on more discussion of the dependencies (lines 129 and 130). However, Fruitloops repeatedly asks how they can construct a square (line 131). They have constructed many triangles in previous sessions, but never a square.

There is again considerable experimentation taking place in GeoGebra during **Log 41**. Note from the time stamps that this log spans over 20 minutes. The three students take turns trying various approaches using the tools they are familiar with and gradually adding the perpendicular line tool. Again, there is a period of intense exploration of the tools and how they can be used. This exploration is not only similar in its intensity to previous periods when they experimented with the GeoGebra tools in preparation for constructing a perpendicular bisector or an inscribed triangle. It also incorporates some of the tool-usage practices they developed then and supplements them with the use of the GeoGebra perpendicular-line tool. They are increasing their tool understanding, which is necessary (as mediator) for understanding the problem of constructing a square.

Log 41. The team constructs its first square.

131	38:45.9	fruitloops	how but how do we make the square?
132	39:11.5	cheerios	a grid
133	39:11.6	fruitloops	like i know how to make the triangle but now the square
134	39:16.5	cheerios	a grid
135	39:20.3	cornflakes	olets start by cinstructing a regular square
136	39:48.0	fruitloops	i think we should make perpendicular lines somehow
137	39:58.8	cheerios	use the perpindicular line tool
138	43:21.9	fruitloops	the first line segment would be like ab
139	43:27.7	cornflakes	yes
142	51:24.7	cheerios	how do u know ji is straight
143	55:40.6	fruitloops	i dont know what to do because the points arent the same color
144	56:38.2	fruitloops	now after you make the perpendicular lines try to make the circles\
145	57:48.7	fruitloops	i think you need to know use the polygon tool and make the square
146	59:10.6	fruitloops	now we need to use the compass tool lilke we did in the triangles tab

147	59:57.5	fruitloops	because af is equal to ec and dh and bc
148	00:42.4	cheerios	i made a line segment which was if than i used the perpendicular line tool and made 2 lines on each side then used the compass tool and clicked on each point and then the center vertex was i and then made a another circle except the center vertex is j and connected all the points

The team is considering the definition of a square as having all right angles, so they first talk about using a grid and then constructing perpendiculars. In line 143, Fruitloops questions how to construct the square in such a way that the points are the same colors as in the original inscribed-squares figure. She sees the fact that her tentative construction has different colored points than the example figure as a sign that there is a problem with her attempt.

Finally, Cheerios succeeds in constructing a dynamic square (see **Figure 29**), and describes the procedure in line 148. The student construction of the square is quite elegant. It closely mirrors, parallels and builds upon the construction of an equilateral triangle, which the students have mastered and the team has adopted as a group dependency-related practice. The square has a base side (segment IJ) and two circles of radius IJ centered on I and J (like the triangle construction). For the right-angle vertices at the ends of the base, perpendiculars are constructed at I and J. Because segments JK and IL are radii of the same circles as IJ, all three segments are constrained to be equal length (by the same reasoning as for the three sides of an equilateral triangle). This determines the four vertices of a quadrilateral, IJKL, which is dynamically constrained to be a square. Although we can see this justification in the procedure of the construction, the students do not spell this out in their chat.

> *Group dependency-related practice #11: Construct a square with two perpendiculars to the base with lengths dependent on the length of the base.*

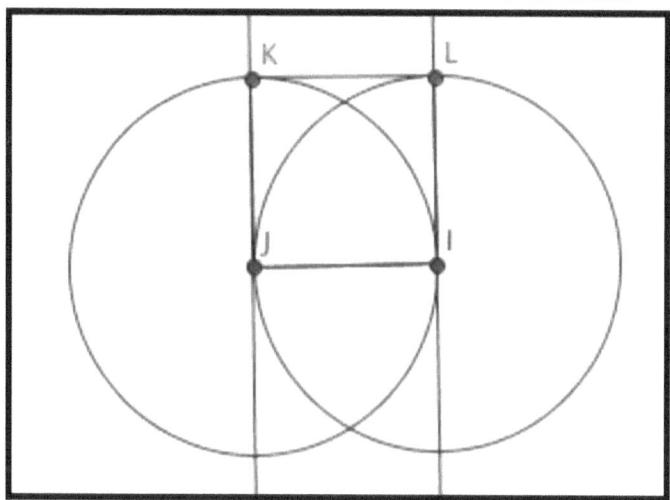

Figure 29. The team constructs a square.

Again, Cheerios surprises us. In the previous session, Cheerios (rather than Fruitloops, who one might have expected) was the one with the insight into the relationships among segments between the two triangles in the example figure. Cheerios (rather than Cornflakes) was the one who proposed the construction procedure. Now, Cheerios (rather than Cornflakes) is the one who actually constructs the square. Cheerios has come a long way from the early topics, in which she often seemed confused or at a loss about what was going on or what to do. How did she succeed in the innovative achievement of constructing a square? Again, in the details of the group interaction, we can observe some of the sharing of individual skills that takes place in this collaboration.

The problem solving does not proceed smoothly, but begins with a number of false starts. When the group begins to discuss making a square, Cheerios first proposes using the grid display, which provides a grid of squares across the screen (lines 132 and 134). The others ignore this proposal. Cornflakes calls for a construction of a "regular square" and Fruitloops suggests, "i think we should make perpendicular lines somehow" (line 136). Cheerios immediately responds positively to this suggestion and tries to adopt it: "use the perpindicular line tool."

Following up on this, Cheerios repeatedly tries to use the perpendicular tool, but fails to create perpendicular lines. The tool use requires that one select an existing line or segment, but Cheerios only selects points. She creates a segment, but then does not construct a perpendicular to it. No one helps her. Instead, Fruitloops and Cornflakes try using circles. Fruitloops succeeds in constructing a quadrilateral connecting intersection points on circles, but the sides are visibly not perpendicular to each other. Nevertheless, this may have provided an image for Cheerios.

Cheerios asks Fruitloops, "how do u know ji is straight" (line 142). Cheerios probably meant, how does one know that segment IJ is precisely horizontal. This suggests that Cheerios is still partially oriented to the visual prototype, as though only a horizontal line can have a perpendicular to it. When she gains control, Cheerios nevertheless builds on segment IJ, erasing everything else that Fruitloops had created. She successfully uses GeoGebra's perpendicular tool to construct perpendiculars to segment IJ at its endpoints, I and J. She then uses the compass tool to construct circles centered on I and J, with radii of length IJ—applying the group construction practice established by Fruitloops. As she does this, Fruitloops types, "now after you make the perpendicular lines try to make the circles" (line 144). The students are watching each other and providing advice and encouragement.

Cornflakes takes control briefly, but does not contribute to the construction. Then Cheerios constructs the remaining vertices (K and L) for the square at the intersections of the circles with the perpendiculars (see **Figure 29**), while Fruitloops is typing "i think you need to know use the polygon tool and make the square" (line 145). In conclusion, Cheerios has been successful through careful attention and perseverance: she worked hard to get the perpendicular tool to work for her and she followed Fruitloops' approach of building circles around segment IJ (as in the equilateral triangle construction of the previous session). Fruitloops' chat postings show an implicit group coherence, even though the postings were too late to have guided the actions.

As soon as the outer square is constructed, Fruitloops proposes to construct an inscribed inner square by following a procedure analogous to the procedure they used for inscribing the triangle. While she narrates, the team actually constructs the inscribed square and conducts the drag test on it. The team's speed and unanimity in taking this step demonstrates how well they had learned the lesson of the inscribed triangles.

> *Group dependency-related practice #12: Construct an inscribed square using the compass tool to make distances on the four sides dependent on each other.*

In lines 146 and 147, Fruitloops proposes using the compass tool to construct the interior, inscribed square, because they have to maintain dependencies of equal lengths from the vertices of the exterior square to the corresponding vertex of the interior square. They used the compass tool in the last tab to maintain analogous dependencies in the triangles. In line 146, Fruitloops states, "now we need to use the compass tool lilke we did in the triangles tab." Note the use of the plural subject, "we," referring to the team and proposing an action plan for the team—based on what the team did in the previous session. Fruitloops is "bridging" back to past team action as relevant to the current situation of the team (Sarmiento & Stahl, 2008a). In line 147, she continues to draw the analogy between the line segments in the inscribed squares with those of the inscribed triangles.

The students bridged back to a practice they had adopted in the previous session: group dependency-related practice #6: Construct an inscribed triangle using the compass tool to make distances to the three vertices dependent on each other. This was a new practice for the group, which had not yet become automatic, habitual or tacit in their behavior. Furthermore, it had to be adapted from the context of triangles to that of squares, so it needed to be brought into the explicit discourse of the group. However, since it had already been adopted by the group in an analogous context, it was easy for the group to quickly apply it here.

Although the chat log (**Log 42**) is dominated by Fruitloops, review of the dynamic-geometry construction using the Replayer shows that the construction of the inscribed square is again a team accomplishment. Note that there is no chat posting for the crucial minute and a half while the inscribed square is constructed between postings 153 and 154. During this interval, Cheerios picks up on the plan and creates point M at 02:27. Cheerios continues to create point O at 03:36, point N at 04:02 and point P at 04:16. Fruitloops immediately comments approvingly of this construction act (line 154).

Log 42. The team makes another key observation.

149	01:07.5	fruitloops	correct
150	01:15.3	cheerios	then used to polygon tool and then hid the circles and lines
151	01:36.9	fruitloops	and we used the circles to make the sides equal because the sides are their radius
152	02:39.8	fruitloops	point m is like point e because it moves around
153	02:48.8	fruitloops	and its the same color
154	04:14.4	fruitloops	good!!

While Cheerios does most of the construction of the inscribed squares, everyone on the team takes turns in control of the GeoGebra tools and contributes to the process, displaying in various ways that they are paying attention and supporting the joint effort. From 02:21 to 04:43, Cheerios constructs the figure, as shown in **Figure 30**. At 04:52, Cornflakes steps in and hides the circles made by the compass tool to define the lengths of the segments along the four sides as equal—just as Cheerios and Fruitloops had discussed in chat lines 146 and 148. Following this, both Cheerios and Fruitloops perform the drag test to check that their new figure preserves its dependencies of inscribed vertices and equal sides. Cheerios drags point M starting at about 05:30 and then Fruitloops drags points I and M starting at about 08:00.

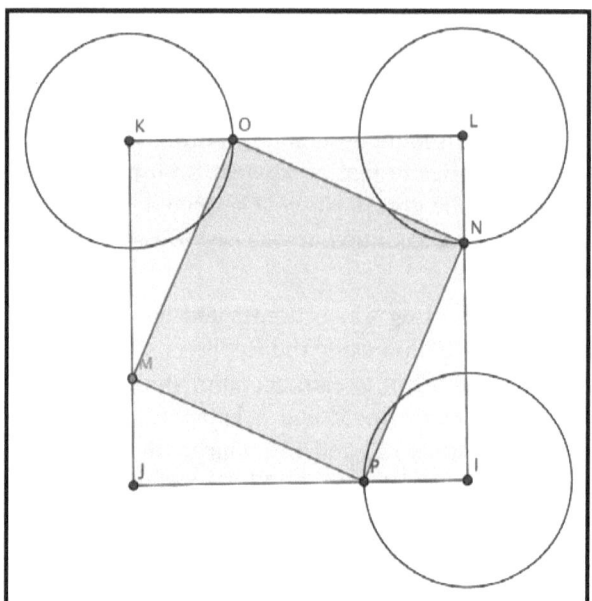

Figure 30. The team's inscribed squares.

In **Log 43**, the team hides the construction circles that impose the necessary dependencies. They use the drag test and conclude that their construction works the same with the circles from the compass tool being hidden or invisible as it did with the original, visible circles (see **Figure 31**).

> *Group dependency-related practice #13: Use the drag test to check constructions for invariants.*

Log 43. The team tests its construction.

155	04:40.4	fruitloops	now hide the circles
156	05:25.7	fruitloops	the points match up
157	05:47.2	cheerios	yay it works
158	06:00.8	fruitloops	it works! just like the original circl;e

159	08:23.8	cornflakes	yess

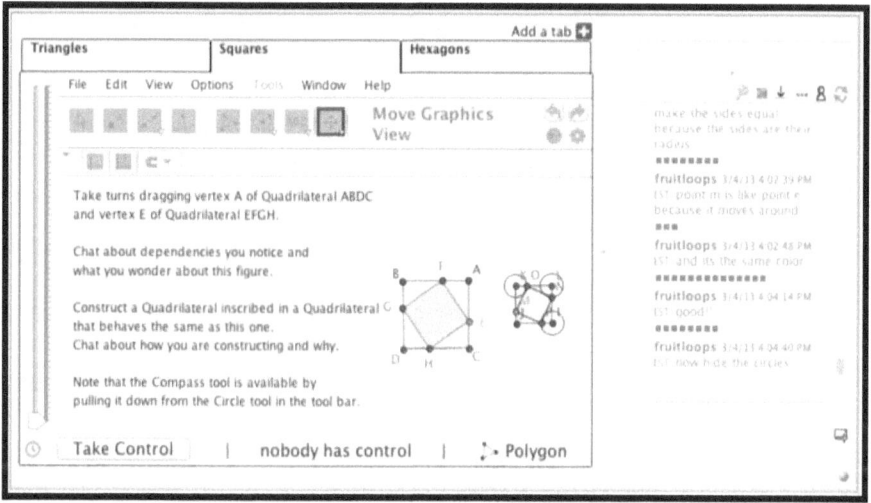

Figure 31. The team constructs an inscribed square.

In line 159, we see Cornflakes responding approvingly to the result of the drag tests and to Fruitloops' conclusion about the dependencies introduced by the compass tool's circles. The team expresses its general agreement with their accomplishment (lines 157 to 159), displaying their shared understanding of their group work. They celebrate that their construction works. It was a challenging task and the team accomplished it. They understood what they were doing—based on other construction practices (line 158, where "circl;e" was probably meant to be "square"). They have verified that it works and is therefore successful.

Cheerios summarizes the procedure they followed in **Log 44**. Fruitloops notes (line 160) that the points N, O and P are colored dark because they are completely dependent upon where the compass circles intersected the sides of the exterior square (originally, before they were hidden). Cheerios reiterates (line 162) that it is the distance along the sides up to these points that is constrained to be equal and Fruitloops agrees, clarifying that the distances are constrained because they are all dependent on the same radii. Cornflakes agrees. Although not all the manipulation of the compass tool is visible to the other team members, enough of Cheerios' work in GeoGebra—especially the sequencing of her construction—was shared on all the computers to display meaningfully to the others the significance of that work. It was not just the visual appearance of the final figure that was important here, but also the design process of the construction of dependencies with compass circles. The circles could be subsequently hidden, so that they were not visible in the final appearance. However, the students were aware of the constraints that remained at work.

Log 44. The team summarizes dependencies.

160	09:18.8	fruitloops	i think points o, n, and p are dark because they weere made using the original circles
161	09:32.4	cornflakes	yea i agreeee
162	09:42.5	cheerios	so just plotted a random point on line segment jk and then used the compass tool and clicked on point m and j (radius) and then clicked k to be the center and then plotted the point where line segment kl intersect with the circle and repeated these steps on the other sides
163	10:02.9	fruitloops	yeah i saw and i understand
164	10:18.5	cornflakes	same
165	10:27.8	cheerios	the distance between m and j is the same between ok and ln and pi
166	10:45.0	fruitloops	all the radii are the same so the distances from ko,ln, and ip and jm are the same
167	10:57.3	cornflakes	yup i agrree
168	11:01.3	cheerios	yes
169	11:03.5	fruitloops	should we move on?
170	11:34.0	fruitloops	actually i dont think we have enough time
171	12:02.8	cheerios	yeah so next time

Fruitloops notes (line 166 in **Log 44**) that the segments between the outer and inner square along the sides of the outer square (IP, JM, KO, LN) are constrained to be equal in length because they are dependent on circles that were constructed with the compass tool to have radii that will always be equal to the length of JM. Point M moves freely on JK just like point E on AC, and M is the same color as E, indicating that it has the same degree of constraint. The other students concur and bring the session to a positive conclusion. They have completed the assignment in the second tab and are out of time before they can start on the third tab.

The chat in **Log 44** is confusing because some of the postings overlapped in their typing, so that some lines respond to postings other than the immediately preceding one. We have to check the full log to reconstruct the threading of responses. Cheerios took two and a half minutes (from 07:10 to 09:42) to type up line 162, carefully documenting her construction steps. In line 165 (09:45 to 10:27), she continued this description, explaining that the construction created equal line segments. Strikingly, Fruitloops typed an almost identical posting, in line 166 from 10:03 to 10:45. This displays an impressive degree of alignment. Cornflakes immediately (10:52 to 10:57) posts line 167, displaying her agreement as well.

Intertwined with the preceding thread are several others. First, Cornflakes' "yess" in line 159 is probably an aligning response to the antecedent drag testing by Cheerios and Fruitloops. Second, Fruitloops responds to line 162 in line 163, stating that she saw and understood the construction steps that Cheerios now describes. Cornflakes then joins in by saying "same" in line 164, indicating that she too saw and understood the construction sequence. In addition, Cornflakes agrees in line 167 to Fruitloops' claim about dependencies in line 166 and Cheerios agrees with Fruitloops statement in line 166, which was so similar to Cheerios' own statement in line 165. The need to involve threading relationships and to understand postings as responses to preceding events or as elicitations of future events is indicative of analysis at the group-cognitive unit.

The excerpt in **Log 44** displays a high level of agreement among the three participants. Often the actual mathematical problem solving or geometric construction is done jointly by the team, with two or three of the participants taking turns doing the steps. However, even when only one person does the actions, the others are intimately involved in planning the moves, describing them or evaluating them. Each major action is discussed and the team agrees to its correctness before moving on to another task. Generally, each action by an individual is entirely embedded in the group context and situated in the team interaction. Geometry construction acts make sense in terms of team plans in the preceding chat and/or team reflections in the subsequent chat. Individual chat postings make sense as responses to preceding actions or comments. The three students in this study repeatedly display for each other (and indirectly for us as analysts) that their activity is a team effort. Through their repeated agreements and other group practices, they constitute their activity as such a team effort.

This is a nice example of group cognition. It contrasts with the pejorative sense of the term "group-think" in mass psychology, which involves a thoughtless form of acceptance of authority or irrational conformity. The analysis of this session shows that thinking as a group involves considerable cognitive effort on the part of each individual involved and is by no means automatic, easy or common. Yet it is possible, given proper guidance and a supportive collaboration environment. And it is the foundation of mathematical cognitive development for the team members.

Summary of Learning in Session 6

In their session on inscribed squares, the team is expanding from their previous session on their ability to construct figures with dependencies. Moreover, this approach is shared by all the members of the team. No one is talking much about visual appearances in this session. The team's work in this session solidifies their development in the previous session. In particular, we have identified five additional group dependency-related practices adopted by the team in this session:

9.   *Points in GeoGebra are colored differently if they are free, restricted or dependent.*

10. *Indications of dependency imply the existence of constructions (such as regular circles or compass circles) that maintain the dependencies, even if the construction objects are hidden.*

11. *Construct a square with two perpendiculars to the base with lengths dependent on the length of the base.*

12. *Construct an inscribed square using the compass tool to make distances on the four sides dependent on each other.*

13. *Use the drag test routinely to check constructions for invariants.*

Combined with the group dependency-related practices identified in the previous chapter, these additional practices form a substantial group understanding of how to construct figures with dependency relationships. They enabled the team to construct both inscribed triangles and inscribed squares. The construction of inscribed triangles is a significant challenge, which is rarely achieved within an hour even by adults with considerable mathematical experience. The team's almost instantaneous construction of the inscribed squares once they had created the outside square demonstrated how well the students had learned the lessons of the triangles—that is, how well the group had adopted a set of practices involved in the earlier group interaction.

At the end of this session of the Cereal Team, we see how the team improves:

i.   Its *collaboration* has expanded to the point that everyone seems to generally be able to follow what everyone else is doing in GeoGebra (unless they are exploring how to use certain tools or trying out an approach that has not been discussed). They take turns almost automatically and they reach mutual understanding and agreement rapidly. There are still some long periods without chat, and that could certainly be improved upon. The team of three students has worked quite closely throughout the two-hour double session. They have collaborated on all the work, taking turns to engage in the dragging and construction. Together, they have discussed the dependencies both in the original figures and in their re-constructions. They have tried to ensure that everyone on the team understands the findings from the dragging, the procedures in the constructions and the significance of the dependencies. In their first hour working on Topic 5, the team successfully constructs an inscribed equilateral triangle using what they had previously learned. In the second hour, they figure out on their own how to construct a square. They also succeed in the task of re-creating the inscribed square. Most of this work is accomplished *collaboratively*. However, for some of the exploration—such as how to construct a square—the students work on their own without much communication. (Presumably, they watch each other's work even then, since only one could have control of GeoGebra at a time.) However, it may not have always been easy to tell why certain actions were being conducted. The rest of the time, the team works collaboratively: each member explains what she is doing during the key GeoGebra actions, everyone confirms that they understand each step and they take turns with the steps so that the major accomplishments are done by the group as a whole.

ii.   Its productive mathematical *discourse* is quite effective. The team is able to plan construction approaches, guide whoever has control and describe the significance of what they have done. Once a student figures out something, she shares it with the others. Not only do they describe what they do during their productive periods and provide some insightful reflections on why their solution is valid, but they also demonstrate a firm grasp of the insights into the solution procedure by immediately applying the same procedures to construct the inscribed square as they had discovered for the inscribed triangle. Their productive mathematical discourse is limited to making sure that everyone understands the basic ideas, without necessarily spelling them out explicitly using mathematical terminology.

iii.   Its ability to use the *tools* of GeoGebra is improving markedly. In both the Triangles Tab and the Squares Tab, the team begins with exploratory dragging to get a sense of dependencies in the original figure. Then they experiment with constructions, guided by some sense of what to look for and things to try, but without a clear plan or an explicit strategy. Eventually, they discover a good solution, describe it explicitly, test it with the drag test and discuss the underlying dependencies that make it work. In particular, the team now seems to understand when the compass tool is appropriate and they are able to use the compass tool—which most people find tricky to learn—effectively.

iv.   Its ability to identify and construct dynamic-geometric *dependencies* has improved dramatically. This group-cognitive mathematical development is accomplished through the continued use of the group dependency-related practices from the previous session and their expansion in this session. The team certainly now identifies different kinds of dependencies in example figures well. It often associates this with possible constraint mechanisms, such as confining specific points to compass circles. The students discuss relationships among geometric objects in terms of *restrictions, constraints and dependencies*—sighting a number of forms of evidence. In the preceding analysis of the earlier sessions, we saw that the three students became quite aware of the different dependency status of certain free points (points A and B), constrained points on lines (point D) and dependent points at intersections (points C, E and F). They had learned that these different statuses are indicated by different coloring of the points in GeoGebra, and they were concerned to make the points in their re-created figure match in color the corresponding points in the original figure. They explicitly discussed points placed on a line being *constrained* to that line during dragging and points defined by intersections (of two circles, of two lines or of a circle and a line) being *dependent* on the intersecting lines and therefore not able to be dragged independently. As they work on the tasks in this session and discuss their findings, the group develops a more refined sense of dependencies. One can see this especially in the way that one student restates another student's articulation of dependencies and how everyone in the group agrees to the restatement.

Their skill with the fundamentals of dynamic geometry has certainly improved. In particular:

a.   *Dynamic dragging*: the whole team practices vigorous dragging to identify what is and is not constrained, both in the example figures and in their own constructions. They finally begin to use the drag test routinely to check the maintenance of invariances through constructed dependencies. All three students adopt the drag test as a regular practice.

b. *Dynamic construction*: the discovery of the construction of a square is a major accomplishment of this session. The team had not been given any instruction in this. They invent a method that is for them a mathematical discovery. Their solution is elegant and they justify it well, if not with a complete explicit proof.

c. *Dynamic dependency* in GeoGebra: the team displays in their solution of the inscribed-squares task a depth of understanding of their solution to the previous inscribed-triangles task. This involves designing a construction to maintain specific dependencies among elements of a figure. They all immediately see that the inscribed squares requires the same kind of dependency relationships as the inscribed triangles.

It is hard to draw implications for changing this set of tabs, except to allow a full hour for each tab. The Cereal Team had impressive success and made significant progress during this double session. The teacher did some extra preparation: showing the compass-tool video in class, discussing dependencies and printing out the tab instructions on paper. The teacher also reminded the students to discuss in the chat their GeoGebra plans and actions. It would have helped the group to chat more during their long periods of GeoGebra exploration—they could have helped each other and shared what they learned better. This suggests additional guidance that could be given for these tasks and others.

Session 7: The Team Uses Transformation Tools

The Cereal Team's teacher skipped from Topic 5 to Topic 8 in order to give her students an experience with rigid transformations. Transformations provide a different approach to middle-school geometry (Sinclair, 2008), included in the Common Core curriculum (CCSSI, 2011) and supported by GeoGebra. Transformations create new kinds of dependencies among dynamic-geometry objects, which can be explored through construction and dragging. As we shall see, while the students did get a first introductory experience with transformations through this topic, the one session was not adequate for covering the whole set of different transformations illustrated in even the first tab. The experience with transformations was not focused enough to convey a clear sense of the radically different paradigm of dependency implicit in GeoGebra's transformation tools.

The Transformations Tab

The Transformations Tab (**Figure 32**)—the only tab the students worked on in their seventh session—includes an example of rigid transformations. An original triangle ABC is twice *translated* in the direction and the distance corresponding to vector DE. This produces triangles $A'_1B'_1C'_1$ and FGH, which are dependent upon triangle ABC and upon vector DE for their shape and positions. Then triangle FGH was *rotated* around point I three times, creating three more triangles, all dependent upon triangle FGH for their positions.

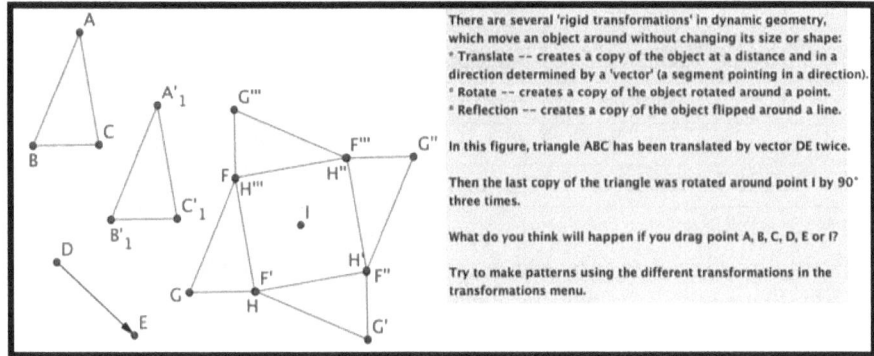

Figure 32. The Transformations Tab.

Cheerios is the first to see what happens when certain objects are dragged (line 8 in **Log 45**). She drags triangle ABC around quite a bit, but does not drag its individual vertices. She observes that all the other triangles move around in response to the movements of ABC. Cornflakes then takes control and drags point A and then point D of vector DE. She sees that changing the shape of triangle ABC by dragging point A changes the shapes of all the other triangles correspondingly (line 16). She also sees that dragging point D changes the positions of all the triangles—both their distance from triangle ABC and their angle from it (line 20).

Log 45. The team drags triangle ABC.

3	11:50.1	cheerios	hey
5	12:32.6	fruitloops	hey
6	12:38.4	cheerios	hello
7	13:18.9	cornflakes	hi
8	14:21.7	cheerios	if u click on triangle abc everything moves it contro;s everything
9	14:51.2	cornflakes	wich abc? theres two of them
10	15:03.1	fruitloops	what happens if you try to move the other triangles/
11	15:09.5	cheerios	abc on the outside
12	15:23.4	cheerios	without the ones
13	15:24.0	cornflakes	theres all on the outside
14	15:33.0	cheerios	number 1
15	15:49.3	cheerios	just click each triangle
16	16:18.0	cornflakes	i am dragging point a and they are all moving!

17	16:29.0	fruitloops	can i try?
18	16:32.6	cheerios	if u click on line segment de it moves too
19	17:09.7	cheerios	they match up with each other
20	17:32.4	cornflakes	yeah they all move the same amount of degrees
21	17:49.7	cheerios	yea and vertex i is the center point

Fruitloops takes her turn and drags point A and points D and E. Cheerios remarks that point I defines the center point around which a number of the triangles move (line 21) as the vector is manipulated. Fruitloops distorts the triangles by dragging the vertex of triangle ABC and notes all the other triangles "move to be the same size and shape as triangle abc" (line 30 in **Log 46**). She also reports that none of the points of the dependent triangles can be dragged directly (lines 34 and 35).

Log 46. The team drags vector DE.

30	18:35.2	fruitloops	the other shapes all move to be the same size and shape as triangle abc only when you move abd
31	18:46.9	cheerios	abd?
34	18:56.2	fruitloops	all the other points can move by themselves
35	19:00.9	fruitloops	cant**
36	19:04.2	cornflakes	yes
37	19:27.0	cheerios	so what now
38	19:33.1	fruitloops	and the line de controls in what position the traingles are but it doesnt affect their shape
39	20:29.0	cornflakes	we can try to make patterns using different transformations in the transformations menu?
40	20:42.6	fruitloops	also when you move de it doesnt affect triangle abc

Fruitloops completes her observations by stating that the vector determines the locations of the transformed triangles, but not their shapes (which are determined by the original triangle) (line 38), and that the vector does not affect the original triangle (line 40). Cornflakes takes control and checks this. She wonders why changing the vector does not affect triangle ABC: "why thoough?" (line 41 in **Log 47**) and "pls find out" (line 43). Note that here Cornflakes adopts the kind of questioning that was initially characteristic of Fruitloops' chat contributions. Even the precise wording that Fruitloops used in early sessions --"why thoough?"—is repeated.

Log 47. The team questions transformations.

41	20:59.1	cornflakes	why thoough?
42	21:13.8	fruitloops	i do not know
43	21:24.3	cheerios	pls find out
44	22:12.7	cheerios	please*
45	22:57.6	cheerios	the instructions
46	23:11.2	cornflakes	all the triangles are in some sort of formation except for triangle abc
47	23:15.1	fruitloops	do we have to try to make this?
48	23:30.2	fruitloops	what about i?
49	23:35.4	cornflakes	yes we have to make patterns

The students have rather systematically used dragging to investigate the dependencies of the example figures. They have nicely summarized their observations of how the size, shape and locations of the transformed triangles are controlled by the locations of the vertices of triangle ABC and the endpoints of vector DE. However, they do not seem to grasp the notion of transformations, as expressed in the instructions. In contrast to the instructions, they did not try to predict what would happen when the different points were dragged, but immediately went about dragging them. Even after the team has analyzed all the movements, none of the students can articulate any understanding of why the triangles move the way they do as a result of the construction described in the tab instructions (lines 41, 42, 43). Cheerios references the instructions in line 45, but there is no discussion of what they mean.

For instance, the instructions define "Translate – creates a copy of the object at a distance and in a direction determined by a 'vector' (a segment pointing in a direction)." If one reads this carefully and thinks about the causal relationship implied by it, then it should be clear that the location of the copy of the object will vary with changes of the vector, but the location of the original object will not. Apparently, the students do not read this carefully or are not able to mentally visualize the consequences of the relationships. Rather, they need to explore the relationships visually in the GeoGebra graphics. Perhaps as they come to understand what takes place visually on their computer screens they will eventually be able to represent it mentally. That is the ultimate goal.

The team's remarks on the results of dragging are interspersed with suggestions that the group move on to the final step of making a pattern. While they have observed the visual patterns of the transformed triangles as the original points are dragged, they do not understand the relationships that cause these coordinated movements well enough to plan a new arrangement of shapes and transformations that will result in a similar pattern of moving shapes.

Cheerios explores the dragging more, extensively dragging the points of triangle ABC and vector DE. Fruitloops notices point I, which no one has tried to drag, and she asks Cheerios (who currently has control) to drag it (line 50 in **Log 48**). Cheerios drags point I all around.

Log 48. The team finds the transformation tools.

50	23:49.8	fruitloops	cheerios try dragging i...
51	25:35.8	cheerios	i ddi
52	26:06.2	cheerios	did*
53	27:00.0	cheerios	guys say what you are doing
54	27:20.9	fruitloops	i only affects triangles ghf g"f"h" and g"'h"'f"'
65	33:04.5	fruitloops	does the circle compass tool have to do with this?
66	33:52.7	cornflakes	it says transformations menu
67	34:06.0	cornflakes	yea
68	37:12.9	cheerios	what do we do
69	37:27.3	cornflakes	can i take control?
70	37:35.4	cheerios	yea
71	37:51.7	fruitloops	i dont know what to do
72	38:04.1	cheerios	same
73	39:14.9	cornflakes	its the third to last box
74	39:26.5	cheerios	?
75	39:35.2	cheerios	the tool?
76	39:36.5	cornflakes	the transformation menu box thing
77	39:40.5	cornflakes	yeas
78	39:40.8	cheerios	oh okay
79	39:44.7	fruitloops	i see but i dont know how to use them
80	40:01.6	fruitloops	anyone want to tyr?
81	40:12.4	cornflakes	in dont either
82	40:19.7	cornflakes	cheerios can u try
83	40:28.6	fruitloops	yeah cheerios
84	40:48.4	cheerios	i dont know what to do sorry

Fruitloops takes control next. She drags points A and B around, without stating in the chat what in particular she is exploring (line 53). She just mentions that point "i only affects triangles ghf g"f"h" and g"'h"'f"'" (line 54). This observation is not tied to the statement in the instructions, which says that a triangle was rotated around point I three times.

After the teams' extensive drag tests, the students look around for a couple of minutes, including at the other tabs. Finally, Fruitloops tries to understand the construction process that created the dependencies the team has been observing. She wonders, "does the circle compass tool have to do with this?" (line 65). Until now, complex dependencies that the team has explored have involved the compass tool or the circle tool.

The team tries to understand the problem given in this topic. But understanding of a geometry problem and of how to approach its solution is mediated by ones practical knowledge of available tools for manipulating and constructing figures like those involved in the problem. The Cereal Team has no prior knowledge about the transformation tools that are involved in the Topic 8 figures. In particular, the GeoGebra transformation tools use a paradigm that is unfamiliar and confusing until one gets used to it. When a given figure is transformed, the original figure remains on the screen and its transformed version is added. This is different from other tools like dragging (a kind of translation): when a given figure is dragged, only the resultant figure remains. The team tries to figure out how the transformation tools work by dragging the topic's example figures, but the GeoGebra paradigm of its transformation tools is not apparent to them. Thus, they do not have the required background tool-usage understanding to mediate an adequate comprehension of the problem, let alone conjectures about how to approach a solution. Fruitloops tries to use her knowledge about the compass tools to substitute for knowledge of the transformation tools.

Although Cornflakes objects that the instructions refer to the transformations menu rather than the compass tool for constructing the figures, Fruitloops explores whether she can get similar results with the compass tool. She creates a triangle JKL (**Figure 33**) with the polygon tool. Then she constructs a circle of radius equal to the length of side KL about point M with the compass tool. She places a point N on the circumference of the circle and then constructs a ray MN through it. She changes this to a vector MN. Fruitloops drags M and N, seeing that they do not affect triangle JKL. Then she drags point K of the triangle and sees that this changes the size of the circle. She concludes her experiment by saying, "i dont know what to do" (line 71). Apparently, Fruitloops wanted her vector to control the size or location of the triangle, but she got the causality backwards: her triangle side length controlled the vector length MN (or circle size), which was dependent upon side KL.

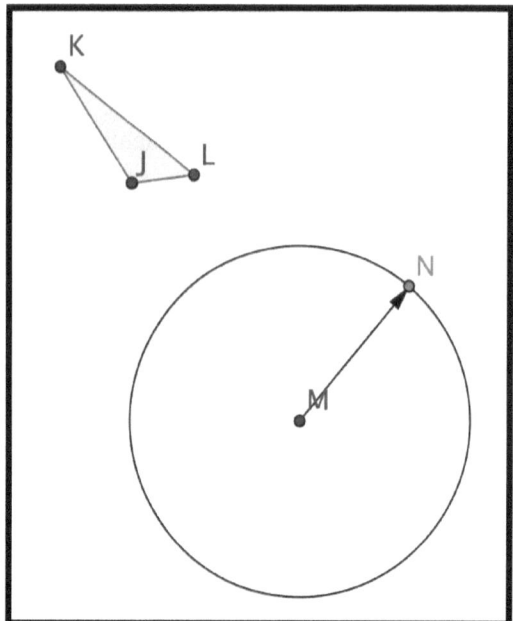

Figure 33. Compass with vector.

Meanwhile, Cornflakes locates the transformations menu in the GeoGebra tool bar and shares that information. Fruitloops says, "i see but i dont know how to use them" (line 79). The students try to get each other to start with these new tools and Cornflakes eventually tries (line 85 in **Log 49**). She creates two triangles with the polygon tool and twice selects the transformation tool, Reflect Object About Line. She says "i am trying to create triangkles and make them function in a patttern" (line 88), but she does not know how to make use of the transformation tool.

Log 49. The team reflects a triangle.

85	41:39.9	fruitloops	corn u try
86	42:14.0	cheerios	did u get it
87	42:44.1	fruitloops	cornflakes what are you doing?
88	43:15.2	cornflakes	i am trying to create triangkles and make them function in a patttern
89	43:30.5	fruitloops	how? what tool are you using?

90	46:14.7	fruitloops	i reflect triangle qrs about line tu but idk how to make it pattern
91	48:53.9	cheerios	i dont see anything
92	49:31.3	cheerios	explain daniella
93	49:47.1	cheerios	fruitloops*
94	50:12.0	fruitloops	so i first mad triangle qrs and then made a verticle line right next to it
95	50:56.2	fruitloops	i used the "reflect object about line tool" and made triangle q'r's' \
96	51:08.7	cornflakes	how did u get the other triangles
97	51:13.2	fruitloops	so q'r's' is just a reflection of qrs
98	51:20.9	cheerios	yea im a little confused
99	51:29.8	fruitloops	then i made a horizontal line underneath both those triangl;es
100	51:42.2	cornflakes	and then youn reflected t?
101	51:46.0	cornflakes	i understand now
102	51:50.6	cheerios	ohhh i see
103	52:14.5	fruitloops	i again used the reflecting tool and reflected qrs to make r'1 s'1q'1 and s"r"q"
104	52:26.3	fruitloops	you should try opther tools

Fruitloops takes control and tries out the tool that Cornflakes selected. She first creates a triangle with the Rigid Polygon tool, perhaps because this topic is concerned with "rigid transformations." She immediately deletes it and replaces it with triangle QRS (see **Figure 34**). Then she creates line TU when she realizes she needs a line to reflect the triangle about. At first, when she selects the Reflect tool, she clicks where she thinks the reflected triangle should go, simply creating new isolated points. Then she successfully uses the tool (possibly based on the displayed tool help) by clicking on her triangle and the line. She announces that she was able to reflect the triangle, but that she still does not know how to make a pattern of triangles (line 90).

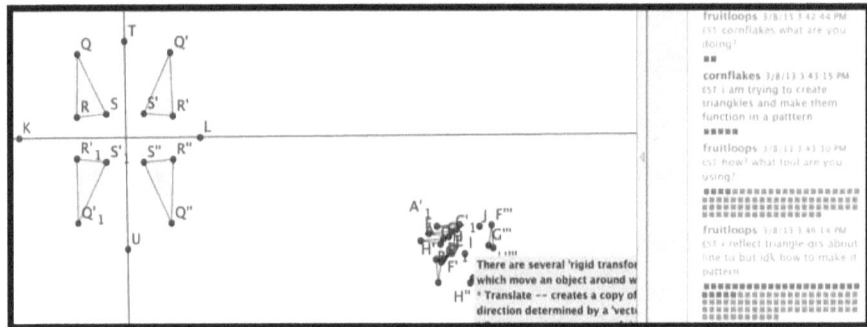

Figure 34. Reflections of a triangle.

Fruitloops did her construction off to the side of the workspace to avoid confusion with existing points. Unfortunately, Cheerios does not see this on her screen and cannot follow what is going on from line 91 to line 102. Meanwhile, Fruitloops places line KL roughly perpendicular to line TU and reflects her two triangles about the new line, forming a pattern similar to the tab's original example, although using a different procedure. She drags point S and sees the other triangles change shape in a coordinated way, like a choreographed pattern. She hides (but does not delete) the lines of rotation (line 107 in **Log 50**) and drags point S more. Fruitloops explains what she did and both Cornflakes and Cheerios catch up. Fruitloops then suggests that the others try different transformation tools.

Log 50. The team constructs more transformed objects.

105	53:06.8	cornflakes	okay ill go
106	53:15.2	fruitloops	and then i hid the other lines
107	53:33.0	fruitloops	i hid the line si reflected over*
108	54:08.6	cheerios	why are some of the points black and some blue
109	54:21.8	fruitloops	in what?
110	54:51.5	cheerios	nvm cornflakes made a new shape
111	55:28.3	cheerios	i dotn know what to do
112	57:05.2	cornflakes	fruitloops can u try again??
113	00:17.8	cornflakes	shall we move on?
114	00:24.9	cheerios	yes we shall
116	00:30.3	fruitloops	i think so

Cornflakes creates a triangle MNO and makes a brief attempt to transform it (line 110). Then Cheerios tries to reflect the triangle about a point (line 111).

Fruitloops then takes control and successfully reflects Cornflakes' triangle about a point V, creating a new reflected triangle (see **Figure 35**). Then she reflects the new triangle about the same point, overlaying the original triangle with a third one. She continues back and forth, layering reflected triangles on top of each other. When she finishes, Cornflakes drags a point around, watching the two triangles move together.

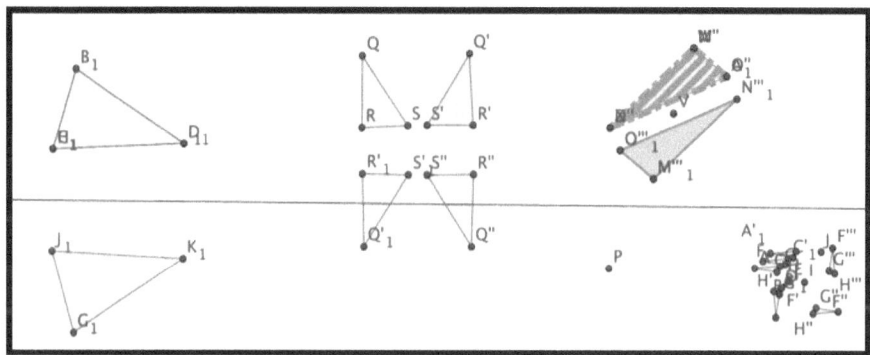

Figure 35. Another reflection of a triangle.

Cornflakes and Fruitloops spend the next two minutes changing the color and appearance of the triangles and their sides (see the striped triangle in the upper right of **Figure 35**). The team decides to move on to the next tab (lines 113, 114, 116). However, as they start to change to another tab their teacher enters the chat room and greets them (line 130 of **Log 51**). She is particularly concerned whether each student has been able to use the transformation tools and asks, "have you guys practiced the transformations?" (line 135). Cheerios and Cornflakes respond, "Yes" (lines 136 and 137). When the teacher asks "which ones?" Cornflakes and Fruitloops respond more specifically. The teacher compliments them (line 145), using the VMT chat pointing tool to reference Fruitloops' answer (line 142). Cheerios thanks the teacher, which prompts the teacher to reference Cheerios (line 146) and ask her, "did you try any tools?" (line147).

Log 51. The teacher visits the chat room.

130	03:11.6	emilyL	hi girls
132	03:23.0	cornflakes	hello
133	03:29.7	cheerios	hey
134	03:35.0	emilyL	I see you have found the icon to change the interior of the polygons
135	03:54.7	emilyL	have you guys practiced the transformations?

136	03:59.4	cheerios	yes
137	04:04.6	cornflakes	Yes
138	04:09.0	emilyL	which ones?
140	04:26.9	cornflakes	the reflect object acrooss line
141	04:33.8	fruitloops	i used the reflect object about line and reflect object about point tools
142	04:57.5	fruitloops	and i made two seperate transformations
143	04:59.7	emilyL	did any of the other group member try different tools as well?
144	05:05.7	cornflakes	i did
145	05:06.7	emilyL	very good job
146	05:16.3	cheerios	thank you
147	05:36.2	emilyL	did you try any tools?
148	06:05.3	cheerios	i tried but i couldnt get it
149	06:10.3	emilyL	try again please
152	06:16.4	emilyL	your teammates will help you
153	06:28.2	cheerios	help!!
156	06:45.4	fruitloops	i tried to explain how i used the reflective tool
157	06:51.6	cornflakes	so the third to last box contains the transformations
158	06:55.9	cheerios	i dont get it though
159	07:08.3	cornflakes	make a polygon
160	07:11.9	cheerios	ok
161	07:15.2	emilyL	explain what you don't get cheerios
162	07:34.7	emilyL	I'll leave you guys to finish and help your teammate. good luck!

The teacher encourages Cheerios to try some more. She says, "your teammates will help you" (line 152). As Fruitloops and Cornflakes start to guide Cheerios, the teacher exits (line 162). They ask Cheerios what transformation tool she wants to use and she says "reflect object about line" (line 168 in **Log 52**). They instruct her to start with a polygon and a line (not a line segment).

Log 52. The team helps each other.

| 163 | 07:37.4 | fruitloops | cheerios, what tool are you trying to use? |
| 164 | 07:45.9 | cheerios | i just made a polygon |

165	08:04.4	cheerios	what now
166	08:07.4	fruitloops	but what transformation tool are you trying out?
168	08:54.2	cheerios	reflect object about line
169	09:07.2	cheerios	its the first one
170	09:10.6	cornflakes	so i think you have to make a line first.....
171	09:19.7	cheerios	where next to it?
172	09:22.0	fruitloops	okay so first you make a polygon and then you make a line next to it
173	09:34.4	cheerios	done
174	09:38.9	fruitloops	line not line segment
175	09:50.9	cornflakes	i said line
176	09:57.5	cheerios	done
177	10:02.3	fruitloops	actually i dont know if it makes a difference but i did it with a line
178	10:08.9	cornflakes	so did i
179	10:15.0	cheerios	okay ill just stick to that then
180	11:17.7	fruitloops	they you select the refelct object about line tool and then clicki on your polygon and then thje line you are reflectting over and another obvject should appear
181	11:58.8	cornflakes	yes
182	12:33.8	cheerios	i got it
183	12:35.2	fruitloops	yay!
184	12:38.7	cornflakes	yes!!
185	13:10.5	fruitloops	you can try the object about point tool cause its very similar

When Cheerios has her triangle and line, Fruitloops gives her detailed instructions for using the reflection tool: "they you select the refelct object about line tool and then clicki on your polygon and then thje line you are reflectting over and another obvject should appear" (line 180). It takes Cheerios about a half a minute. First, she selects the reflection tool but then clicks on points where she expects the new triangle to appear. After deleting her new points, she clicks on a side of her triangle, making the triangle reflect about its side rather than about the line. Finally, she succeeds and proclaims, "i got it" (line 182). Fruitloops and Cornflakes celebrate with her. Fruitloops then suggests Cheerios try a similar transformation tool, but the time is over for the session.

Summary of Learning in Session 7

In their session, the team has a first experience with GeoGebra's tools for rigid transformations. The team does not adopt any major new practices as a group.

At the end of this session of the Cereal Team, we see how the team improves:

i. Its *collaboration*: the three worked well together to explore the transformation tools. Cornflakes and Fruitloops managed to locate and use certain of GeoGebra's transformation tools. Cheerios could not completely follow this or duplicate it. So the other two guided her— at the suggestion of the teacher, who intervened to make sure that each student could use the tools.

ii. Its productive mathematical *discourse*: the team does not have very good discussions of how the transformation tools work.

iii. Its use of GeoGebra *tools*: the team gains a first exposure to the set of transformation tools, which embody a different paradigm of dependency and dynamic movement.

iv. Its ability to identify and construct dynamic-geometric *dependencies*: while the students can identify dependencies through dragging, they do not form clear conceptions of how the dependencies work with transformations.

We also track the team's fluency with:

a. *Dynamic dragging*: the students do not understand the causality that is displayed by their dragging.

b. *Dynamic construction*: the team has some preliminary success in constructing transformations. They would need more sessions to start to understand the paradigm.

c. *Dynamic dependency*: the group has begun to explore the dynamic dependency created with transformations and rotations.

The students' introduction to transformations was not well scaffolded. The topic was presented out of sequence, with insufficient preparation. The teacher apparently just wanted the students to try GeoGebra's transformation tools, not necessarily to understand them as a radically different approach to dependency than the use of circles or compass tool in the previous topics. However, the problem presented by the topic was too complicated. To understand it, one already had to know something about how the transformation tools worked. The students should first explore simple examples of using one transformation tool at a time.

It was already noted in the introduction to this chapter that the group did not really gain an understanding of the new geometric paradigm involved with transformations, which historically developed well after Euclid. The task given to them assumed too much prior understanding of what transformations are. The curriculum should be revised to introduce transformations gradually, with simple examples and more explanation. In particular, the topic of transformations should be tied to the exploration of dependency in dynamic geometry. Transformations represent a different notion of dependency or a set of new mechanisms and tools for implementing dependencies. Most of Euclid's examples of dependency used circles (or their equivalent in the GeoGebra compass tool). In addition, there are simple dependencies constructed by placing a point on a line or at the intersection of two lines. Transformations open up a completely new world of dependencies, which are more complicated to understand, at least as implemented in GeoGebra.

Session 8: The Team Develops Mathematical Discourse and Action Practices

In their final session as part of WinterFest 2013, the Cereal Team displays considerable growth from how they began in their early sessions. Topic 13 (worked on in Session 8) presents a large number of quadrilaterals with different dependencies. The team has quite different degrees of success in identifying the dependencies of the first several quadrilaterals, which they investigate. Note that with the exception of the Inscribed Squares Tab, this is the group's first encounter with quadrilaterals, as opposed to triangles.

In our research project, we are particularly interested in seeing how well the group understands how to design dependencies into dynamic-geometry constructions. For Topic 13, the instructions begin by asking, "Can you tell how each of these quadrilaterals was constructed? What are its dependencies?" During the one-hour session, the team posted 174 chat messages discussing in order the first seven of the displayed four-sided figures, Poly1 through Poly7. In this chapter, we will follow the interaction of the team during this session to see how well the group could identify the dependencies in each of the different quadrilaterals and how well it could surmise how they were constructed.

In our presentation, we will build on the more detailed analysis of dragging as a referential resource for mathematical meaning making in this session by Çakir and Stahl (2015). We will especially use Çakir's graphical representations of key dragging sequences.

Taking a somewhat different approach here, we will highlight a series of *group mathematical practices of discourse* (text chat) *and action* (GeoGebra dragging and construction) that the Cereal Team adopted during this session. In dynamic geometry, many of the most important mathematical action practices involve some form of dragging (Hölzl, 1996) and/or construction. This topic does not include the students engaging in active construction, so we will primarily see the students adopting group practices of dragging. However, the topic does explicitly involve reflection of how the given figures were constructed. This highlights the mediation by ones understanding of potential construction tools for ones understanding of geometric figures.

As this was the team's final session, our analysis will serve as a partial summative evaluation of the extent of the team's development of mathematical group cognition. This will involve the identification of group practices of mathematical discourse about dependencies, since the focus of the curriculum is on understanding dependencies in dynamic geometry.

The Quadrilateral Tab

The team begins this session immediately and efficiently, with confidence. Although the screen is covered with a potentially bewildering array of labeled quadrilaterals, the team begins to drag the points of the first quadrilateral without hesitation. Each student takes a turn manipulating its four vertices and discusses what she observes while doing so. They also note the appearance (esp. color) of the points at the vertices. They quickly conclude that all four vertices of Poly1 are unrestricted in their movement. They check that the team members are all in agreement. They discuss the vertices in terms of dependencies (or lack of such). They use geometric terminology and labeling. They even describe how the quadrilateral must have been constructed to have the observed behavior. In contrast to their earlier behavior, they now attribute the observed characteristics of the geometric figure to the construction process as the source of dependencies. Thus, as we shall see, the group demonstrates in the opening minutes of their final session a level of collaborative exploration, mathematical discourse, familiarity with construction tools and reflection on constructed dependencies far more advanced than in their early sessions.

The VMT interface is shown in **Figure 36**. This image taken from the VMT Replayer shows the task for Topic 13. Specifically, it shows an important moment in the final session, which corresponds to **Log 56** below at about line 73. The group is dragging points E, F, G and H of Poly2 in the GeoGebra tab named "Quadrilaterals" while they are chatting in the chat panel. Topic instructions are included with the 22 pre-constructed dynamic-geometry quadrilaterals in the GeoGebra tab.

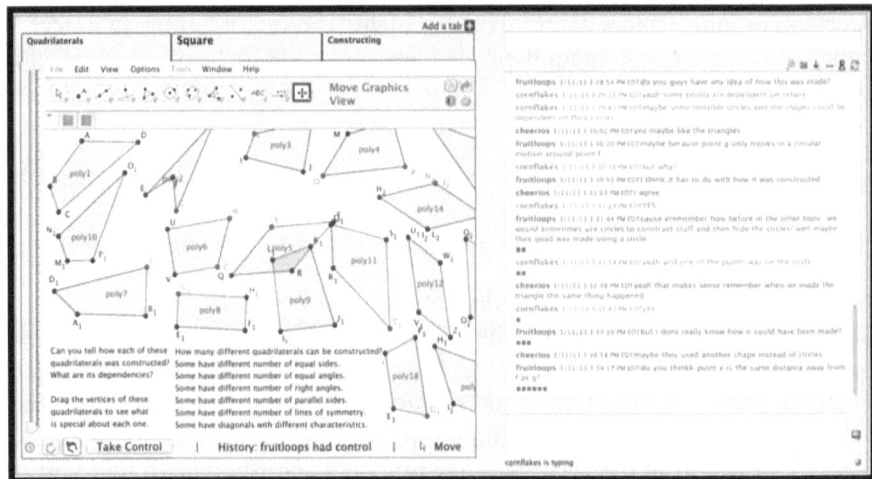

Figure 36. The Quadrilateral Tab.

As they go through the session, however, the students seem to just identify which points are restricted. They do not go on to identify the kinds of quadrilaterals that are defined thereby. In other words, they do not necessarily display a clear sense that the figures were designed to have certain properties (like equality of 2 or 3 or 4 side lengths) and that these were implemented by constructing dependencies with the hidden circles that they suspect are somehow at work. Thus, the question we need to investigate in the analysis of this session is the extent to which the team has gotten the main intended message of the VMT WinterFest about designing the construction of dependencies in order to impose desired dynamic behavior on the resultant figures. To do so, we review the interaction about each of the polygons in order, as the group took them up.

Poly1: Efficient Analysis

The team's discussion of Poly1 is impressively straightforward and efficient— especially when contrasted to their interactions in their first sessions, a couple weeks earlier. The three students enter the room and they each take a brief look at each of the three tabs before beginning to interact (lines 1-12, not shown in the log). Then Fruitloops proposes starting by looking at the first example of a quadrilateral, Poly1, which is labeled ABCD (see line 13 in **Log 53**).

> *Group mathematical practice #1:*
> *Identify a specific figure for analysis.*

Log 53. The team analyzes Poly1.

13	15:00.2	fruitloops	lets start with quad abcd
14	15:18.5	fruitloops	in the upper lefthand corner
15	15:47.4	cornflakes	ok
16	16:20.1	cheerios	label it by saying its points
17	16:26.5	fruitloops	okay so for poly 1 all the points can move anywhere and i dont think they have resrictions
18	16:42.3	cornflakes	ok
19	17:19.1	fruitloops	so i think this was constructed by just making four points and using a polygon tool
20	17:38.1	fruitloops	you guys can try moving if youd like
21	18:14.2	cornflakes	yeah your right i dont think theres any restrictions
22	18:23.2	cheerios	can i try
23	19:00.7	cheerios	there are no restrictions like you said
24	19:34.6	fruitloops	so do you agree with how i think it was constructed
25	19:38.6	cornflakes	yes
26	19:44.8	cheerios	yes
27	19:53.4	fruitloops	okay good

Fruitloops opens the chat with a post that initiates the discussion of Poly1: "lets start with quad abcd." She directs her teammates' attention to it by referencing its name ("poly 1" in line 17), vertex labels ("quad abcd" in line 13) and position ("in the upper lefthand corner" in line 14) in the displayed GeoGebra tab. Cheerios seconds the use of point labels: "label it by saying its points." The use of point labels for referencing initiates a group practice of indexicality originally defined by the ancient Greek geometry community and now adopted by the Cereal Group in their context:

> *Group mathematical practice #2:*
> *Reference a geometric object by the letters labeling its vertices or defining points.*

Fruitloops drags each of the vertices and sees that each one moves independently. She drags point A twice and each of the others just once before announcing, "okay so for poly 1 all the points can move anywhere and i dont think they have resrictions" (line 17).

While waiting for responses by her teammates, Fruitloops drags the vertices some more and concludes, "so i think this was constructed by just making four points and using a polygon tool" (line 19) Cornflakes and Cheerios agree to Fruitloops' proposal to start with Poly1. She encourages her partners to try moving the vertices for themselves, and they do so. Then they affirm both her observation about a lack of restrictions on the movement of the vertices and her proposal of how the figure may have been constructed. The team then moves on to the next figure.

They have followed the several steps of the instructions in the tab for Poly1: dragging each vertex, determining dependencies (or lack of them) among the figure's components and suggesting how the figure could have been constructed. Furthermore, they have all taken turns dragging and agreeing to each conclusion.

Poly1 is the simple, base case of a quadrilateral with no special relationships among its sides or angles. Therefore, its construction is a trivial application of the generic polygon tool of GeoGebra. The tools involved are well understood by the team. Led here by Fruitloops, the team is incredibly efficient at: focusing on the task of their new topic; exploring the geometric figure's dynamic behavior; concluding about the lack of dependencies; proposing how the figure was constructed; having everyone in the team explore the figure; having everyone agree with the conclusion; having everyone agree with the construction proposal; and then moving on to the next task. In going through this process, the team established a multi-step approach that they could then follow for each of the subsequent polygons.

Fruitloops, as an individual, did the original exploration by dragging, proposed the solution to the task and led the group through it. Because of the simplicity of the task for an individual with the experience that Fruitloops now has, there was no need for group cognition or group agency in this case. Nevertheless, if one compares this chat excerpt with the log of the team's first session, the episode demonstrates that this particular team of three students has learned a lot about collaborating and interacting in the VMT environment, using the GeoGebra tools, enacting the practices of dynamic geometry and engaging in problem solving.

Poly2: Group Memory

The discussion of Poly2 is particularly complex to analyze, especially in contrast to that of Poly1. There are overlaps in the typing of chat postings, leading to multiple threads of discourse. In particular, conceptual interchanges about the meaning of terms like "constrained" and "dependent" are mixed with practical explorations of the constraints among geometric objects in the GeoGebra tab.

Let us look at the opening lines of the team chat about Poly2 (**Log 54**). Cornflakes volunteers to take the lead with this polygon. As shown in **Figure 37**, Cornflakes drags vertex F in a counter-clockwise direction around point E.

Log 54. The team explores constraints in Poly2.

28	19:58.3	cornflakes	ill go next?
29	20:13.2	fruitloops	sure
30	20:26.5	cornflakes	ill do polygon efgh
31	20:37.5	cheerios	just say the number its easier
32	21:17.3	cornflakes	okasy polygon 2 has all points moving except point g
33	21:28.8	cornflakes	and point g is also a different color
34	21:40.3	cheerios	do u think it is restricted
35	21:44.7	cheerios	or constrained
36	21:49.4	fruitloops	i feel like poly 1 and poly 2 are almost exactly the same except that poly 2 had one point that is a lighter shade
37	22:04.5	fruitloops	can i try moving it?
38	22:17.1	cornflakes	sure
39	22:25.0	fruitloops	and @ cheerios , i dont know for sure
40	23:18.0	cheerios	ok can i try
41	23:22.7	cornflakes	sure
42	23:23.3	fruitloops	so point g only moves in like a circular motion around point f
43	23:35.6	cornflakes	@fruitloops yea
44	24:16.7	cheerios	what si the difference between constrained and restricted
45	24:24.3	cheerios	is*
46	24:41.6	cornflakes	constrained is limited function

47	24:46.4	fruitloops	also when you move e, g moves away or closer to f
48	25:08.4	fruitloops	so i think g it definitly constrained
49	25:14.0	cornflakes	yes
50	25:19.8	cornflakes	i think that too
51	25:25.6	cheerios	why though
52	25:59.3	fruitloops	and g moves whenever you move point e and f but it doesnt move when you move h
53	26:20.3	cheerios	okay
54	26:42.4	fruitloops	@ cheerios. i think its constrained because it moves but the function is limited
55	27:36.8	cheerios	oh i see
56	27:37.5	fruitloops	what is the definition of dependant
57	28:52.4	cheerios	u need the other line or point otherwise it wont work

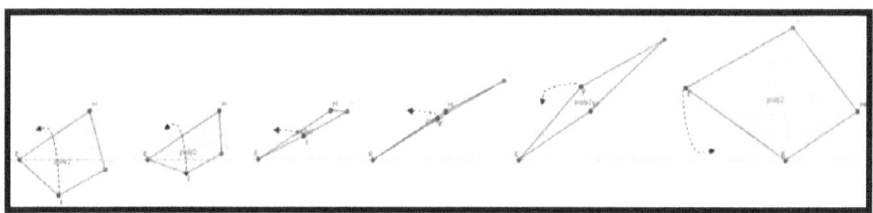

Figure 37. Cornflakes drags vertex F of Poly2.

Each of the six momentary screen-shots corresponds to a view of the polygon as Cornflakes drags it. The added dashed line at the level of point E (which was not affected by dragging F) is provided to aid the visual comparison of successive views. Added small arrows indicate the direction of the dragging of vertex F.

This form of dragging takes a particular—possibly representative—static view of a figure and varies it without violating the dependency restraints imposed on it as a dynamic-geometry construction. This allows the viewer to consider the range of possible configurations that the figure can take on. In classical Euclidean geometry, the mathematician conducted an analogous variation mentally in imagination; here, students who have not yet developed such mental skills for geometric figures can drag the representations in a dynamic-geometry environment to observe the range of views. This is an important practice, which the team adopts:

> *Group mathematical practice #3:*
> *Vary a figure to expand the generality*
> *of observations to a range of*
> *variations.*

Cornflakes subsequently drags each of the vertices systematically in order to explore how the different vertices move. She finds that points E, F and H move freely, but point G does not (line 32). She also notes (line 33) that point G is a different color than the other vertices, which indicates a different degree of dependency in GeoGebra. Another practice involving dragging is adopted here: exploring conjectures about invariances:

> *Group mathematical practice #4:*
> *Drag vertices to explore what*
> *relationships are invariant when*
> *objects are moved, rotated, extended.*

Cheerios asks if this means that point G is "restricted" or "constrained" (lines 34 and 35). This initiates a new thread of discussion about the meaning of the terms "constrained" and "restricted" (line 44).

Meanwhile, Fruitloops requests control of the GeoGebra tab; she drags point G extensively and then point E as well—for about 14 minutes, from line 40 to line 76. Cheerios asks to have control (line 40), but never really takes over control from Fruitloops and remains focused on discussing the issues of constraints—both the definition of the term and its application to Poly2. This sets up two parallel threads of discussion, which both elaborate on Cornflakes' initial observation in line 32.

Fruitloops first explores point G, which Cornflakes had said did not move. (She probably meant that G did not move freely or independently, which is what they were supposed to determine for the vertices). Fruitloops drags point G extensively (see **Figure 38**). She notes that point G's movement is confined to a circle around point F (as long as points E, F and H remain fixed): "so point g only moves in like a circular motion around point f" (line 42).

> *Group mathematical practice #5:*
> *Drag vertices to explore what objects*
> *are dependent upon the positions of*
> *other objects.*

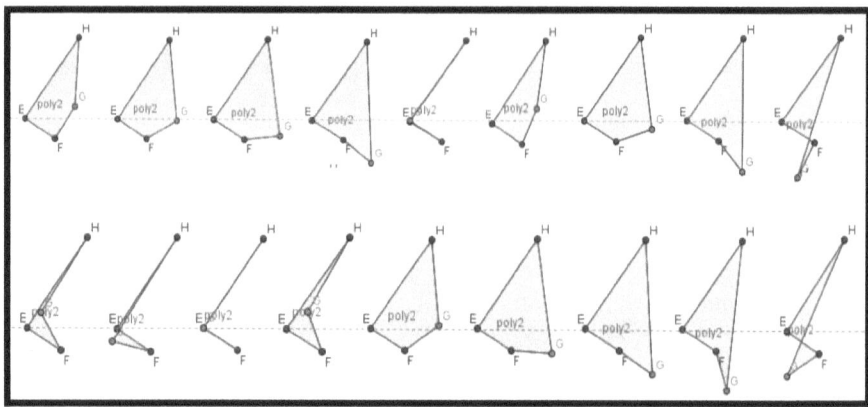

Figure 38. Fruitloops drags vertex G around point F.

Fruitloops then drags point E and discovers that the position of point G shifts in response to movements of point E, changing the length of segment FG: "also when you move e, g moves away or closer to f" (line 47). Relating her findings to Cheerios' discussion of constraint, Fruitloops concludes that G is definitely constrained (line 48). Specifically, G moves in response to changes in E or F, but not in response to changes in the position of H (line 52). Through her extensive and systematic exploratory dragging, Fruitloops has identified the following regularities in the dynamic diagram:

a) G moves around F in a circle, and when G is moved no other vertex moves.

b) When H is moved, no other vertex moves.

c) G moves when F is moved.

d) G moves when E is moved.

e) E and G are always equidistant from F.

> *Group mathematical practice #6: Notice interesting behaviors of mathematical objects.*

We can see this sequence of exploration and noticings as similar to the analysis of Poly1, in which Fruitloops builds on her own postings to accomplish the task of determining the dependencies in the figure. With Poly2, Cornflakes and Cheerios follow Fruitloops' discoveries and express agreement with Fruitloops' conclusions (lines 38, 40, 43, 49, 51, 53).

In parallel, Cheerios asks for terminological clarification in line 44: "what si the difference between constrained and restricted." Cornflakes responds that being constrained means having a limited function (line 46). Fruitloops then provides her analysis of point G as an example of a constrained point, because its ability to be dragged is limited by the positions of other points (E and F): "so i think g it definitly constrained" (line 48). Cornflakes agrees with that in lines 49 and 50. The team is refining its shared understanding of the central terms of the VMT curriculum: constraint, restriction and dependency.

> *Group mathematical practice #7: Use precise mathematical terminology to describe objects and their behaviors.*

Cheerios questions this example by asking "why though" (line 51). This question may seem ambiguous. However, Fruitloops treats it as asking how her analysis of point G fits the definition of constrained that Cornflakes had offered. Extending her conclusion in line 48 that G is constrained (line 52), Fruitloops then adds a remark (line 54) explicitly directed to Cheerios and responsive to her question from line 51: "@ cheerios. i think its constrained because it moves but the function is limited." Cheerios expresses satisfaction with Fruitloops' remarks as adequate responses to her question about why point G should be considered constrained. First, she responds, "okay" (line 53) to Fruitloops' summary in line 52 about how point G moves. Then she states, "oh i see" (line 55) to Fruitloops' response to the question in line 54.

While it may be unclear how well Cheerios understands Fruitloops' explanation, it is interesting that Cheerios has assimilated what was in earlier sessions Fruitloops' (and in the previous session, Cornflakes') role of questioning, "why though?" Earlier in the history of the team, Fruitloops would assume the role of raising the theoretical issues by posting this phrase. More recently, both Cornflakes and Cheerios have used this specific question to push the team discourse.

> *Group mathematical practice #8: Discuss observations, conjectures and proposals to clarify and examine them.*

Having clarified Cheerios' query about the meaning of "constrained," Fruitloops then reverses their relative positioning as questioner and clarifier and she asks Cheerios "what is the definition of dependant" (line 56). After a pause of a minute, Cheerios replies, "u need the other line or point otherwise it wont work" (line 57). Although presented as an answer to the question, taken by itself the formulation remains rather ambiguous as a self-contained definition. It includes indexical terms ("the other", "it") whose references are missing and what she means by not working is not well defined. However, her posting does suggest the main idea that the behavior of a particular point is somehow determined by some other point or line. The next excerpt (**Log 55**) will clarify this definition in terms of past experiences of the team as well as the current focus on Poly2.

Log 55. The team discusses the possible construction of Poly2.

58	28:54.6	fruitloops	do you guys have any idea of how this was made?
59	29:15.6	cornflakes	yeah some points are dependent on others
60	29:43.4	cornflakes	maybe some invisible circles and the shapes could be dependent on thos circles
61	30:02.4	cheerios	yea maybe like the triangles
62	30:20.3	fruitloops	maybe because point g only moves in a circular motion around point f
63	30:35.3	cornflakes	but why?
64	30:55.5	fruitloops	i think it has to do with how it was constructed
65	31:03.8	cheerios	i agree
66	31:29.1	cornflakes	YES
67	31:44.4	fruitloops	cause eremember how before in the other topic we would sometimes use circles to construct stuff and then hide the circles? well maybe thiis quad was made using a circle
68	31:58.9	cornflakes	yeah and one of the points was on the circle
69	32:38.1	cheerios	yeah that makes sense remember when we made the triangle the same thing happened
70	32:43.0	cornflakes	yes

While Cheerios is typing her response to Fruitloops' question about the definition of "dependent," Fruitloops raises another question, equally based on the topic description in the tab: "do you guys have any idea of how this was made?" (line 58). Note that the instructions given to the students in the original tab were, "Can you tell how each of these quadrilaterals was constructed? What are its dependencies?" The students have enacted these questions by discussing the definition of the terms "constrained" and "dependent," and by asking how Poly2 was "made." Interestingly, Fruitloops has translated "constructed" as "made," reflecting the fact that the team has not fully understood construction in dynamic geometry as a rigorous mathematical process of defining relationships of dependency, but rather continues to discuss it in informal everyday language as an ill-defined assembling (Khoo & Stahl, 2015).

> *Group mathematical practice #9: Discuss the design of dependencies needed to construct figures with specific invariants.*

Cornflakes responds, bringing together the two threads. First, she affirms and elaborates Cheerios' definition of dependency, stating, "yeah some points are dependent on others" (line 59). Then she responds to Fruitloops' question, using this definition of dependency: "maybe some invisible circles and the shapes could be dependent on thos circles" (line 60). This introduces a discussion by the team that displays their understanding of the role of dependencies in the design of dynamic-geometry figures. Let us see how the team discusses the dependencies designed into Poly2 in **Log 55**.

Consider line 61: Cheerios says, "yea maybe like the triangles." This is a potential pivotal moment in that it initiates a new and productive mathematical discourse direction. It brings in a crucial lesson that the team learned in a previous session about constructing dependencies in triangles. However, it is clearly not a self-contained expression of someone's complete and adequate response to the topic, like Fruitloops' earlier proposal about Poly1. Rather it has the appearance of a semantic fragment, whose meaning is dependent upon its connections to other chat postings.

The first word, "yea," seems to be responding in agreement to a previous statement by another team member. The next word, "maybe," introduces a tentative proposal soliciting a response from others. Finally, "like the triangles" references a previous topic of discussion. Thus, line 61 is dependent for its meaning on its connections to previous postings, to potential future postings and to a topic from another discussion. Line 61 is structured with these various semantic references and the meaning of the posting is a function of its ties to the targets of those references. We will now try to connect line 61 to its references, recognizing that the target postings are also likely to be fragments, dependent for their meaning on yet other postings, ultimately forming a large network of semantic or indexical references.

The "yea" of line 61 registers agreement with line 60, "maybe some invisible circles and the shapes could be dependent on thos circles." Line 61 reaffirms the tentative nature of this joint proposal by repeating line 60's hedge term, "maybe." It thereby further solicits opinions on whether the proposal should be adopted.

Line 61 then adds both detail and evidence in support of the proposal by referencing the lessons that the team experienced in working on "the triangles" in an earlier GeoGebra session. In Session 2, about three weeks earlier, the group had learned how to construct an equilateral triangle by constructing two circles around endpoints A and B of a line segment, both circles with radii of AB. The two circles constrained point C, defined by the intersection of these circles. The fact that the two circles both had the same radius (AB) meant that the sides AC and BC of a triangle ABC (which were also radii of the two circles) would both be equal in length to the base side AB, making triangle ABC always equilateral. So a proposal to take an approach "like the triangles" could involve constructing circles that are later made invisible, but confining new points to those circles to make the figures formed by the new points dependent upon the circles (whether the circles are visible or not) in order to impose equality of specific segment lengths.

The thread from line 61 posted by Cheerios back to line 60 posted by Cornflakes is a response to line 58 posted by Fruitloops: "do you guys have any idea of how this was made?" Line 58 is a call to address the main questions of the session's topic: "Can you tell how each of these quadrilaterals was constructed? What are its dependencies?" When applied to Poly2, it asks how quadrilateral EFGH was constructed, taking into account its dependencies, which the team has been exploring.

So the meaning of line 61 is that it proposes an answer to the topic question as expressed in line 58, building on and confirming the tentative partial response in line 60. The meaning does not inhere in line 61 on its own or on that posting as an expression of Cheerios' mental state, but as a semantic network uniting at least the three postings by the three team members, and therefore only making sense at the group level of the interaction among multiple postings and GeoGebra actions by multiple team members.

The meaning-making network of postings continues with line 62 by Fruitloops: "maybe because point g only moves in a circular motion around point f." Again, this posting begins with "maybe," establishing a parallel structure with lines 60 and 61, aligning or unifying the postings of all three team members. The posting goes on to provide specifics about how the proposed invisible circle could be working, similarly to how it worked for the equilateral triangles. It names point G as the point that moves on the circle and point F as the point at the center of the circle. This is based on Fruitloops' extensive dragging of point G. The posting orients the team to specific points on the screen, in their interaction with one another. It helps the team to see what is going on *as* a certain interaction among those points. It thereby contributes to group geometrical vision.

> *Group mathematical practice #10:*
> *Use discourse to focus joint attention*
> *and to point to visual details.*

The next chat post, by Cornflakes, questions why G would move around F: "but why?" (line 63). This time, Cornflakes has again adopted a phrase that Fruitloops typically used in earlier sessions, identical in form to the "why though" that Cheerios recently adopted (in line 51).

Fruitloops responds to line 63 in detail in lines 64 and 67, tying the observed behavior to the conjecture by Cornflakes and Cheerios in lines 60 and 61 about how Poly2 may have been constructed with a circle. This is based on the team's earlier experience constructing point C of an equilateral triangle on circles and then hiding the circles but having C remain at a distance AB from points A and B. She types: "i think it has to do with how it was constructed" (line 64) and "cause eremember how before in the other topic we would sometimes use circles to construct stuff and then hide the circles? well maybe thiis quad was made using a circle" (line 67). (Note that Fruitloops now uses the term "construct" rather than "make" when she is referring to the step-by-step procedure involving using circles "to construct stuff.")

Lines 65 and 66 from Cheerios and Cornflakes agree with Fruitloops' line 64. Line 68 by Cornflakes then elaborates: "yeah and one of the points was on the circle." This clarifies that not only was it necessary to construct a circle, but then it was necessary to construct one of the points of the quadrilateral on that circle—so it would be constrained to remain on that circle (even if the circle was subsequently hidden from view).

Line 69 sums up this whole discussion relating to the experience from Session 2: "yeah that makes sense remember when we made the triangle the same thing happened." Cornflakes agrees in line 70 with Cheerios' conclusion. Line 69 is a quite explicit and strikingly literal affirmation of successful sense making: "that makes sense ...." It appeals to the team to "remember" the previous experience as directly relevant to their current issue.

The team has effectively bridged from their current task of understanding how Poly2 was constructed back to their past lesson about how to construct an equilateral triangle. The team has—through an effort of remembering (or "bridging" across discontinuous sessions) that involved all three team members working together—recalled relevant aspects of the past shared experience and situated those aspects in the current situation (Sarmiento & Stahl, 2008b). They have made sense of their current problem with the help of their past experience, their previously adopted practices and their habits of tool usage. This excerpt of the chat has displayed for the team and for us evidence of what might be considered group learning or even transfer—and has illustrated certain methods the team used to recall their former experience and tie it to the current joint problem context.

> Group mathematical practice #11:
> Bridge to past related problem
> solutions and situate them in the
> present context.

After posting line 67, Fruitloops resumes her exploration of Poly2. She drags point G, perhaps confirming her posting back in line 62 that "point g only moves in a circular motion around point f," but not stating anything about her observations. Rather, she posts line 71, "but i dont really know how it could have been made?" (See **Log 56**.) This posting destroys the coherence of the team effort. It puts into question the progress the group made without providing any specifics about what the problem might be, let along indicating a path out for group inquiry.

Log 56. The team becomes confused about Poly2.

71	33:10.4	fruitloops	but i dont really know how it could have been made?
72	34:14.5	cheerios	maybe they used another shape instead of circles
73	34:17.4	fruitloops	do you thinkk point e is the same distance away from f as g?
74	35:03.6	cornflakes	we coulda had a shape on a triangle or square made it invisible but in reality the other shape is still there therefore making one of tth e points that was on the shape dependent on that shape
75	35:31.9	cheerios	i think it is the same tool maybe they used the compass tool cuz they have the same distance
76	36:13.9	fruitloops	and h is just completely unrestriced
77	36:30.8	cornflakes	yeah it probably wasnt built on anything
78	36:31.7	cheerios	agreed
79	36:37.6	fruitloops	agreed
80	36:55.4	fruitloops	so h was probably the first point construceted in building the shape
81	37:05.9	cheerios	yeah

The team has come very close to figuring out the construction of Poly2. They have identified the relationship between vertices G and F, namely that "point g only moves in a circular motion around point f." They have also recalled the construction of the equilateral triangle, in which there were "some invisible circles and the shapes could be dependent on thos circles." From this, they have concluded that "maybe thiis quad was made using a circle" and that "one of the points was on the circle." Yet they cannot seem to take the next step, thinking as designers of dynamic-geometry constructions: to propose that a circle be constructed around point F and that point G be placed on that circle.

Cheerios and Cornflakes try to respond to the problem, but their responses do not seem to reflect attention to the GeoGebra dragging of point G that Fruitloops has been doing. Cheerios suggests "maybe they used another shape instead of circles" (line 72). This ignores the apparent circular motion of G around F. Simultaneously, Cornflakes reiterates how the dependency of a point on a line remains even when the line is hidden: "we coulda had a shape on a triangle or square made it invisible but in reality the other shape is still there therefore making one of tth e points that was on the shape dependent on that shape" (line 74). While it is true that the inscribed square or inscribed triangle kept its vertices on the "shape" of the inscribing figure, the dependency that ensured that involved using the compass tool and locating points of intersection between the inscribing figure and the compass' circular shape. Although the hypothesized dependency-producing figure is currently hidden and could therefore in theory have any shape, all of the team's experience with hidden shapes controlling dependencies has involved circles (produced by the circle tool or the compass tool). Thus Cheerios is potentially distracting from the group's insight by suggesting the consideration of other shapes. Cornflakes' comment does not support Cheerios' suggestion.

Fruitloops ignores these postings of the others and asks, "do you thinkk point e is the same distance away from f as g?" (line 73). She then actively drags the points of Poly2 more to explore this conjecture.

> *Group mathematical practice #12: Wonder, conjecture, propose. Use these to guide exploration.*

In particular, she drags point G counter-clockwise (see **Figure 39**). G can be seen as moving in a circle around F. Fruitloops slows down when G is about to move near vertex E. Here, it is clear that point F is the same distance from E as G.

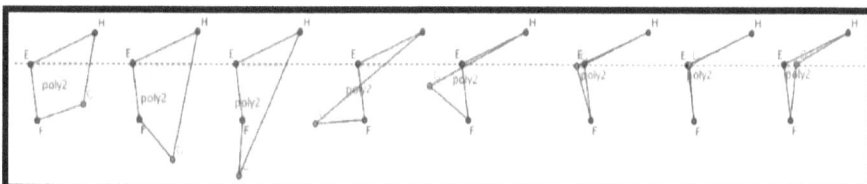

Figure 39. Fruitloops drags vertex G in a circle.

> Group mathematical practice #13:
> Display geometric relationships by
> dragging to reveal and communicate
> complex behaviors.

While Fruitloops is dragging G, Cheerios reverses her previous suggestion and argues for the circle rather than some other shape: "i think it is the same tool maybe they used the compass tool cuz they have the same distance" (line 75). She indicates that "maybe" the construction used the compass tool rather than the circle tool, because the compass also creates a circle and was used for making line segments stay the same length as each other in their recent past experience with the inscribed triangle and square.

Fruitloops continues to ignore the others and produces another proposal: "and h is just completely unrestriced" (line 76). The initial "and" syntactically ties her new posting to her previous one as a continuation, in effect bypassing the postings of the others. Cornflakes and Cheerios quickly agree with the new proposal. Cornflakes draws a consequence of the unrestricted behavior of point H for its construction: "yeah it probably wasnt built on anything" (line 77). Cheerios simply agrees with Fruitloops, posting "agreed" (line 78). Fruitloops posts the identical "agreed" in response to Cornflakes' consequence. The team now seems to be aligned once more. Having pursued an intense discussion with proposals from everyone, the team now resorts to its older pattern of aligning with Fruitloops' explanations.

Fruitloops next proposes a further consequence: "so h was probably the first point construceted in building the shape" (line 80). Cheerios quickly affirms this. However, during the next 6 seconds, there may have been some second thoughts about this. The awareness messages in the chat system indicate that Cheerios started to type another message twice and then deleted it, while Fruitloops also started to type a message that was never posted. Finally, Cornflakes questioned the latest proposal with "whatb do you mean?" (line 82 in **Log 57**). As we shall see from Cornflakes' follow-up postings, this question was intended to open the way for critical considerations. Cheerios then asked if Fruitloops meant "the first point plotted?" (line 83). Cornflakes built on Cheerios' post about the order of constructing points and clarified her non-specific previous question in lines 84 and 85: "doesnt it go in alphabetical order?" and "efg and then h." Without explicitly stating so, these postings presented counter-arguments to Fruitloops' proposal. The group thereby entered into a new practice: considering the succession of steps needed to produce a figure with identified dependencies.

> *Group mathematical practice #14: Design a sequence of construction steps that would result in desired dependencies.*

Log 57. The team is unsure of how Poly2 was constructed.

82	37:14.9	cornflakes	whatb do you mean?
83	37:33.2	cheerios	the first point plotted?
84	37:51.7	cornflakes	doesnt it go in alphabetical order?
85	38:06.7	cornflakes	efg and then h
86	38:16.7	fruitloops	well if h can move anywhere it was probably made first cause if you just put a random point anywhere it is the same dark blue color as h and it can move anywhere
87	38:45.0	cornflakes	but e and f are the same color>>>>>>>>>>
88	39:02.3	fruitloops	but e and f are constrained..
89	39:14.6	fruitloops	i dont know for sure maybe youre right
90	39:57.6	cheerios	im not very sure either
91	40:02.4	cornflakes	meneither

Line 86 from Fruitloops presents an argument for why point H was probably the first point constructed in building Poly2: "well if h can move anywhere it was probably made first cause if you just put a random point anywhere it is the same dark blue color as h and it can move anywhere." The team knows from their previous construction activities that if one simply constructs a point by itself, it appears in the same dark blue color as point H, and one can drag it freely, the way that point H can be dragged. (Recall that in GeoGebra points are colored differently if they are free, constrained or dependent. The students have learned to use this as a clue for determining how a figure may have been constructed.)

Cornflakes points out in line 87, "but e and f are the same color." In other words, E or F could have been constructed before H because they are the same color as H, indicating that they are also free points. Furthermore, they come earlier alphabetically. Fruitloops responds (line 88) that they are not free like H, but can be seen through the dragging that she previously did to be constrained: "but e and f are constrained." Presumably, since the behaviors of points E and F are constrained, they must have been constructed after the free points, like H. However, Fruitloops politely admits that she is not convinced that she is right and that Cornflakes is wrong: "i dont know for sure maybe youre right" (line 89).

Cornflakes then takes control in GeoGebra and drags the vertices of Poly2 extensively for a half a minute. At the end of that, Cheerios concludes, "im not very sure either" (line 90). Cornflakes agrees: "meneither" (line 91). They are confused about what order of creating the points could have resulted in their apparent constraints. Unfortunately, no one attempts to actually construct a polygon with those constraints.

> *Group mathematical practice #15:*
> *Drag to test conjectures.*

This concludes the team's work on Poly2. It seemed that they had figured out the dependencies—that point G maintained a fixed distance from point F and that sides EF and FG were equal, while point H was free. They also seemed close to concluding that Poly2 could be constructed by confining point G to a circle around point F. If they had started to explore such a construction, they would probably have discovered that the circle should have a radius of EF and that would ensure that EF=FG. Unfortunately, the team restricted its explorations to dragging vertices. Of course, this is what the instructions told them to do. They had looked ahead to the instructions for the other tabs and may have seen that trying to construct the quadrilaterals was reserved for the third tab, which they did not have time to work on.

> *Group mathematical practice #16:*
> *Construct a designed figure to test the*
> *design of dependencies.*

The work on this quadrilateral contrasts strongly with that on Poly1. The chat interaction around Poly2 is rich, complex and intertwined. Meaning is created across postings by all three students. Meaning making also incorporates references to the GeoGebra actions, the instructions in the tab, the definitions of key terms, techniques of dynamic geometry and even lessons learned weeks ago. Discussions of the definitions of the terms "restricted," "constrained" and "dependent" are interwoven with observations about relationships between geometric objects. Despite considerable dragging and productive math discourse, the team ends in doubt about its conclusions. Poly2 seems to be a case that is particularly hard to analyze by just dragging; if the team had engaged in construction to explore their ideas about how Poly2 was built, they might have been more successful and confident in their findings.

Poly3: A Confused Attempt

Having agreed that they are not sure how Poly2 was constructed, the team moves on to Poly3, with Cheerios volunteering to be in control of the initial dragging this time (see line 92 in **Log 58**). The others agree (lines 95 and 96).

Log 58. Confusion about Poly3.

92	40:19.3	cheerios	can i do the next polygon
93	40:21.9	fruitloops	should we move on or??
94	40:26.8	cheerios	polygon 3?
95	40:29.9	fruitloops	sure
96	40:34.4	cheerios	alright
97	40:49.6	cornflakes	cheerios your turn
98	40:50.2	cheerios	l is constraned
99	41:14.4	fruitloops	how is l constrained?
100	41:20.4	cheerios	k j l are not restrcited they can move freely
101	41:22.1	cornflakes	yeaH??
114	41:50.0	cheerios	sorry my bad i isnt constrained
115	42:05.8	fruitloops	is l constrained
116	42:08.4	fruitloops	?
117	42:12.8	cheerios	it is l that is constrained
118	42:42.3	cheerios	there is at least one right angle
119	42:42.5	cornflakes	can i get control for a sec?
120	42:49.9	cheerios	sure
121	43:13.0	cornflakes	im not sure
122	43:26.9	fruitloops	i dont really get what you are saying cheerios
123	44:04.5	cheerios	what dont u get
124	44:20.0	cheerios	i dont understand what u mean
126	45:27.5	fruitloops	nevermind'
127	45:43.4	cheerios	okay lol

Cheerios drags point L vigorously and sees that it moves the other vertices, so she says, "l is constraned" (line 98). She may have selected L to explore first because it is colored light blue, like constrained points. Fruitloops has presumably been watching all the movement of the vertices of Poly3 and asks for more detail about how Cheerios thinks that L is constrained, "how is I constrained?" (line 99). Cornflakes reinforces this with "yeaH??" (line 101).

However, Cheerios—who has continued to drag all the vertices of Poly3 as far as possible within the tab—meanwhile revises her analysis repeatedly: "k j l are not restrcited they can move freely" (line 100); "sorry my bad i isnt constrained" (line 114); "it is I that is constrained" (line 117); "there is at least one right angle" (line 118).

Fruitloops asks to be given control of GeoGebra and she drags each of the vertices in many directions. Fruitloops seeks clarification from Cheerios, but it is not forthcoming. After some mutual questioning, they both seem unable to pursue the discussion, erasing their attempts to respond. They mutually agree to move on to the next quadrilateral.

The movements of Poly3 in response to the dragging of a vertex seem quite complex and confusing. Especially if one pulls a vertex a long distance, the whole quadrilateral becomes distorted in strange ways. The problem is that the dependency designed into Poly3 involves sides, not individual vertices. The dependency is that the length of side IJ is equal to the length of side KL (a pair of equal opposite sides). Because any change to the length of IJ will cause side KL to change—while the quadrilateral as a whole has to remain linked up, most attempts to drag any given vertex will cause movements of most of the other vertices. It is a lot harder to see what is going on here than in previous cases. No individual point seems either completely independent or completely dependent on another individual point. It is probably necessary to pose a conjecture about side lengths (like IJ=KL) and then see if it holds up under dragging. Conjectures about individual points do not help. Cheerios' conjecture that "there is at least one right angle" (line 118) also did not pan out.

Poly4: Vertices Swinging around Circles

It is again Fruitloops' turn to drag as the team moves to Poly4 (see **Log 59**). After two minutes of dragging, she determines that "so pont o and p are constrained" (line 129) and more specifically that "point p moves around point n in a circular pattern and o does the same for m" (line 132).

Log 59. Constraints in Poly 4.

128	47:27.3	fruitloops	okay ill do poly 4 now
129	49:36.9	fruitloops	so pont o and p are constrained
130	50:07.7	cheerios	agreed

131	50:15.1	cornflakes	right they are also diff colors
132	50:16.7	fruitloops	point p moves around point n in a circular pattern and o does the same for m
133	50:29.4	cheerios	can i try
134	50:34.3	cornflakes	maybe they were constructed ona circle?
135	50:56.7	fruitloops	maybe
136	51:13.2	cheerios	om and pn are like the radiuses
137	52:16.0	cornflakes	right
138	52:27.6	cheerios	maybe the compass tool?
139	52:40.6	fruitloops	yeah and also when you move point m it changes the distance poitn n is from p and when you move point n it changes the distance between m and o
140	53:03.4	cornflakes	
141	53:11.6	cornflakes	yeah
142	53:32.9	cheerios	yup

Poly4 is apparently easier to analyze. The team can see that points P and O (which are colored as dependent points) swing around points N and M like endpoints of radii of circles. Furthermore, the two radii are connected, so that when you change the length of one that changes the length of the other. The team agrees that this could have been constructed using the compass tool. They then move on.

The team does not remark that when O swings around M, it passes directly over N, indicating that the length of side MO equals the length of side MN. Similarly, NP=MN, so that sides MO, NP and MN are all constrained to be equal by confining O and P to circles of radius MN. The team never addresses the third question in the instructions, to see what is special about each figure—that Poly4 has three equal sides.

Poly5: It's Restricted Dude

Cornflakes tries to drag point T in Poly5 (see **Log 60**) and finds she cannot move it directly. She applies the term "point t is restricted" (line 145). Fruitloops affirms this, citing that point T is colored black, which indicates that it is fully dependent for its position on other objects.

Log 60. A restricted point in Poly 5.

143	53:49.9	cornflakes	oky im going to do polygon 5 now
144	54:33.8	fruitloops	okay
145	54:52.3	cornflakes	point t is restricted
146	55:13.9	fruitloops	agreed because off the color
147	55:33.5	fruitloops	so t only moves when you move the other points
148	55:46.7	cheerios	yea thats one way to prove that is constrained
149	56:09.6	fruitloops	i thought it was restricted
150	56:09.9	cornflakes	and when you move point r all the pointsmove around point q
151	56:29.9	cornflakes	yeah its restricted dude
152	56:48.0	cheerios	sorry that is what i mean
153	57:02.3	fruitloops	okayyy dudeee

Fruitloops then adds, "so t only moves when you move the other points" (line147). Cheerios agrees: "yea thats one way to prove that is constrained" (line 148). Fruitloops questions the use of the term "constrained," saying "i thought it was restricted," which Cornflakes supports: "yeah its restricted dude" (line 151). Cheerios agrees with them that the correct term is "restricted." Point T is not merely partially constrained, for instance to move in a circle maintaining a fixed distance to another point and being constrained to a circular path, but is fully restricted to a specific position relative to other objects.

The team continues to drag Poly5 for several minutes. They drag it into a state where all four vertices are roughly on top of each other. They are not able to drag the vertices apart, but only succeed in dragging labels of the points. So they give up on Poly5 and move on under time pressure.

Poly6: A Rectangle?

The team looks at Poly6 (see **Log 61**). Cheerios drags point Z back and forth a little, ending with Poly6 in a rectangular shape. Cheerios concludes, "z is constrained and it is a square and has 2 sets of parallel lines and has 4 right angles" (line 156). Cornflakes and Fruitloops agree. This is a strange conclusion since the shape does not look completely square. However, it is possible that the students have not learned the distinction between square and rectangle because they have not had a formal course in geometry yet. Actually, Cheerios gives a very nice formal definition of rectangle in terms of what the tab lists as special possible characteristics: having 2 sets of parallel sides and 4 right angles.

Log 61. Parallel lines and right angles in Poly6.

154	01:13.3	fruitloops	lets move on to poly 2
155	01:16.7	fruitloops	6*
156	02:13.6	cheerios	z is constrained and it is a square and has 2 sets of parallel lines and has 4 right angles
157	03:38.2	cornflakes	i agrere
158	03:57.3	fruitloops	i agree*******
159	03:58.9	cheerios	w is constrained also
160	04:05.0	cornflakes	*agree

Still, it is strange that the team accepts this description for Poly6 since it was not rectangular in its original position or all of its other dragged positions. For instance, immediately before announcing that Poly6 (quadrilateral UVWZ) was a square, Cheerios herself had dragged it into the position shown in **Figure 40**. This seems to be a reversion to Cheerios' old way of viewing figures non-dynamically based on their apparent shape at a specific time or their possible shape.

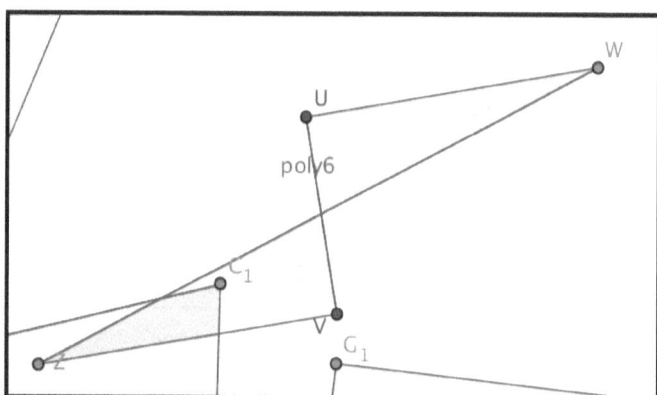

Figure 40. Poly6 in a non-square position.

Perhaps the team (and especially Cheerios) has come up with a different interpretation of "dynamic" figures than the one that is conventional in the dynamic-geometry community. The standard view is that a figure is only called a square if it is restricted to always being a square under dragging. By contrast to this, the team is discussing Poly6 as being a square if it *can* be dragged into a square. Recall that the team had already been exposed to the conventional understanding in Session 2, where they constructed an equilateral triangle that always remains equilateral. However, in Session 4 the students construct a triangle that they consider equilateral despite the fact that it can be dragged into different shapes. They then discuss how they know that it is equilateral because of its measurements (when it is dragged into a square appearance). This alternative view of dynamic figures may be considered a "student misconception"—possibly a useful intermediate representation for making the difficult transition from static visual shapes to dynamic dependencies (Sfard & Cobb, 2014). It is a view that at least some of the students slide in and out of during their work on different problems. Contrast, for instance, Cheerios misconception-based postings here with her impressive work constructing a valid dynamic square and then another square inscribed in it in Session 5. There she was clearly using the conventional understanding of "dynamic." The team has made impressive advances in understanding the dynamic-geometry paradigm, but their grasp of it is not consistent or robust.

Cheerios next drags point W and concludes, "w is constrained also" (line 159). The rest of the team moves on.

Poly7: A Final Attempt

Cornflakes starts to move several of the quadrilaterals out of the way and Fruitloops then moves Poly7 into the cleared space in the tab. After four minutes of silence, she announces "so c1 is deff cpnstricted" (line 161) (see **Log 62**). Cornflakes agrees (line 162) and, when prompted, Cheerios does as well (line 167).

Log 62. A constricted point in Poly7.

161	07:54.9	fruitloops	so c1 is deff cpnstricted
162	08:12.9	cornflakes	yes
163	08:16.1	cornflakes	agreed upon
164	08:19.9	fruitloops	definitly constricted
165	08:55.6	fruitloops	definitely*
166	09:14.2	fruitloops	cheerios do you agrere?
167	09:14.8	cheerios	yea i agrree
168	09:30.5	cheerios	agree
178	10:33.7	fruitloops	sorry

179	10:38.7	cornflakes	soorry
180	11:25.1	fruitloops	that was by accident
181	11:36.3	cheerios	its okay
182	12:05.2	fruitloops	when you mkove a1 c1 also moves
183	12:09.5	cornflakes	yeah
184	12:14.7	cheerios	yeah
185	13:25.7	cornflakes	toodles
187	13:37.2	fruitloops	goodbye fellow peers
188	13:43.5	cheerios	toodles
190	13:55.7	cheerios	nice working with you

Fruitloops continues to test Poly7, mainly by dragging point A_1. She concludes simply that "when you mkove a1 c1 also moves" (line 182). The rest of the log is taken up with repairing typos, apologizing for accidentally sending blank chat messages and saying goodbye at the end of the final session. That ends the Cereal Team's involvement in WinterFest.

Summary of Learning in Session 8

This section has reviewed the work of the Cereal Team during their last of their eight hour-long online sessions of dynamic geometry in the VMT Project's WinterFest 2013. We have analyzed the sequential responses of the team members to each other as the team tries to determine the dependencies in a series of seven quadrilaterals constructed in the VMT dynamic-geometry environment.

In their final session, the team's discourse and actions are more productive than earlier, and they explicitly discuss ideas about dependency. The team works well together, efficiently moving from task to task and collaborating effectively. They take turns leading the explorations of the dynamic-geometry figures, proposing analyses of the dependencies in the figures and deciding when to move on. The team members consistently make sure that they all agree on team conclusions.

The team has varying success in their work on the different figures. With some figures, they are able to make quite complete analyses and come up with reasonable descriptions of how the figures could have been constructed. With other figures, they have much less success.

However, their understanding of construction of dependencies in dynamic geometry is still partial. In a number of ways, the team has not fully enacted the lessons intended by the instructions. The team (primarily Fruitloops) successfully analyzes Poly1, which is an easy case with no dependencies. The team does an impressive job together with Poly2. However, they do not quite succeed in reconstructing the design of its dependencies and they end up in a confusion about the order in which points must have been created. There is not much time left for Poly3 through Poly7, some of which are quite complicated in terms of their dependencies.

The team had varying success in the exploration of dependencies in the seven quadrilaterals they discussed. This variety revealed a significant range in the group's capabilities, from an impressive facility in analyzing dynamic constraints and expressing conjectures about the hidden construction mechanisms to a contrasting inability to see what is going on in other, similar figures:

- Poly1: Fruitloops dragged the figure, noted its lack of dependencies and proposed that it was constructed with a simple use of the polygon tool. The other students took turns dragging the vertices and agreed with Fruitloops. Fruitloops accomplished the task as an individual, and she led the group to a consensus. The collaboration was simple and efficient. The team demonstrated mastery of completing VMT tasks, particularly when their interaction here is compared with that of the early sessions. The learning environment seems to have been successful.

- Poly2: The team worked intensively together on this figure. They brought in many resources, including reflections on constraints and lessons from past sessions. They discussed the concept of dependencies as applicable to Poly2. However, in the end, they were unsure of their findings. It might have helped if they had engaged in exploratory construction.

- Poly3: The relationships in Poly3 were apparently hard to see by just dragging vertices. It might have helped if the team had proposed conjectures, had discussed relationships among sides rather than just between points.

- Poly4: Fruitloops analyzed the dependencies of the vertices. Cornflakes and Cheerios proposed how the quadrilateral was constructed.

- Poly5: The team found a point that is not just constrained to follow a path, but is fully restricted and can only be moved indirectly by dragging another point. The students clarified their understanding of the terms "restricted," "constrained" and "dependent."

- Poly6: Here, Cheerios found that sides are constrained to be parallel. Thereby, they saw that relationships can be among sides as well as vertices.

- Poly7: Fruitloops found a constrained point, but time ran out for the team.

The team was relatively effective in dragging the given figures to explore their dependencies. However, in cases of more complicated dependencies, the team probably did not have sufficient applicable experience in the construction of figures that would result in the observed motions under dragging. For instance, when two sides of a quadrilateral were constrained to be of equal length, the motions could be quite confusing. One would have to make a conjecture about the constraint in order to make sense of the motions with targeted moves. In particular, one would have to imagine using specific GeoGebra tools to impose those particular constraints. The team had not used tools to construct isosceles triangles, isosceles quadrilaterals, kites or parallelograms. So they were not able to use such tool-usage knowledge to mediate their meaning making involving the motions they produced with dragging. More of this construction experience would have been useful either prior to this topic or integrated with it.

In our analysis, we have identified a number of *group mathematical discourse and action practices*, some of them specifically applicable to collaborative dynamic geometry. Most of these occur during the examination of Poly2, which is much richer and more detailed than the team's work on the other quadrilaterals. The identified practices of team interaction using chat and GeoGebra actions are:

1. *Identify a specific figure for analysis.*
2. *Reference a geometric object by the letters labeling its vertices or defining points.*
3. *Vary a figure to expand the generality of observations to a range of variations*
4. *Drag vertices to explore what relationships are invariant when objects are moved, rotated, extended.*
5. *Drag vertices to explore what objects are dependent upon the positions of other objects.*
6. *Notice interesting behaviors of mathematical objects*
7. *Use precise mathematical terminology to describe objects and their behaviors.*
8. *Discuss observations, conjectures and proposals to clarify and examine them*
9. *Discuss the design of dependencies needed to construct figures with specific invariants.*

10. *Use discourse to focus joint attention and to point to visual details*

11. *Bridge to past related experiences and situate them in the present context*

12. *Wonder, conjecture, propose. Use these to guide exploration*

13. *Display geometric relationships by dragging to reveal and communicate complex behaviors*

14. *Design a sequence of construction steps that would result in desired dependencies*

15. *Drag to test conjectures*

16. *Construct a designed figure to test the design of dependencies*

The adoption of these group practices constitute an advance in the team's mathematical group cognition. The display of these practices shows both *that* the team has developed and documents *how* they have developed cognitively— precisely through the adoption of such practices. The combination of these group practices enables the team to engage in a rich discussion of the dependencies involved in dynamic-geometry figures.

At the end of the final session of the Cereal Team, we see how the team has improved during its eight sessions:

i.   Its *collaboration*. The three students move through the tasks efficiently as a team. Sometimes they seem to move too quickly to a next task without coming to an adequate conclusion or reflecting upon their conclusions. They continue to adopt each other's best practices at collaboration. For instance, Cheerios and Cornflakes adopt (e.g., in lines 51 and 63) the probing "but why" questioning that was originally typical of Fruitloops' role in the group.

ii.   Its productive mathematical *discourse*. They consider their understanding of technical terminology in the context of using the terms in their problem-solving work. They have extended discussions about the definition and application of key terms like "restricted," "constrained" and "dependent." They quickly express alignment with each other's proposals, but also engage in critical questioning.

iii.   Its use of GeoGebra *tools*. They make extensive use of various modes of dragging to explore the given quadrilaterals.

iv.   Its ability to identify and construct dynamic-geometric *dependencies* with:

a.   *Dynamic dragging.* Despite considerable use of dragging, their exploration is not always guided by conjectures about the hidden constraints. Thus, they rely too much on the colors of points as clues, rather than interpreting the motions they observe in terms of conjectures about what could cause these behaviors. While colors of points reflect a point's degree of constraint, they do not indicate what the dependencies or dependent relationships are, let alone how they might have been constructed.

b. *Dynamic construction.* The instructions for the tab that they worked on in their final session did not call for construction. However, it would have been helpful to try to construct the dependencies that they proposed, especially Poly2, where they debated the possible order of construction steps.

c. *Dynamic dependency.* The students do not usually connect the types of polygons from the list in the instructions with the constraints they discover, e.g., identifying Poly2 as an isosceles quadrilateral, with two sides of equal length.

There are several implications for redesign of this topic. Clearly, there were too many quadrilaterals displayed in the first tab for a team to deal with in one session, let alone to complete and move on to additional tabs. Consequently, the team rushed through several of the figures without learning much from them. The instructions could have guided the team to formulate conjectures about the hidden dependencies to orient and make sense of their dragging. The instructions could also have emphasized the dynamic nature of the figures— that several might initially look similar, but they could be dragged into very different shapes. Finally, construction could be integrated into the task of the first tab, so that groups could test out their claims about hypothesized dependencies being constructed in certain ways and resulting in the observed dragging motions.

More generally, the analysis in this book—revealing both the remarkable development of mathematical group cognition and the fragility of this understanding—suggests the following implications for the re-design of the VMT environment, especially the curricular resources:

*   The team does not have time to explore all the quadrilaterals or to do any work involving active construction of the figures. This is unfortunate. In previous sessions, the students have also had too little time to do some of the important constructions. For instance, although they did construct an equilateral triangle and a square, they did not construct a simpler isosceles triangle, which might have given them a clearer understanding of the use of circles and the compass tool for imposing the dependency of one segment length on another. Therefore, it would be better to narrow the breadth of topic coverage and focus on a few topics that intensively and clearly involve construction of dependencies.

*   The team seems to be close to a grasp of constraints and dependencies in GeoGebra, but their understand is quite inconsistent and they often have to simply give up on certain figures. It could be quite productive to extend their introduction to dynamic geometry by a couple more sessions. The

group has become very collaborative and efficient, so by the additional sessions they could really focus on dynamic-geometric understanding. A couple more sessions with time-on-task might allow them to become much surer—both as a team and as individuals—of how to explore and construct dynamic-geometry figures with dependencies.

- The student misconception that became particularly clear in this topic should be addressed more clearly in the early topics. The contrast should be explored between (a) figures with *static* visual shapes, (b) figures that *can* take certain shapes and (c) figures that are *necessarily* restricted to certain shapes (which can, however, sometimes be rotated, translated or dilated).

- There should be a clearer presentation—through hands-on problems—of different dependency mechanisms in GeoGebra. The students talk mostly about the use of circles and the compass tool to imposed dependencies. This is natural since many of the tasks used this mechanism. It is also true that Euclid stressed this method in the beginning of his paradigmatic presentation of geometry. However, simply constraining a point to remain on a line or at an intersection is also a dependency mechanism. Defining a custom tool can be seen in terms of imposing dependencies as well. And of course rigid transformations are each dependency mechanisms.

- The team moves from task to task without specific opportunities to reflect on their accomplishments, to compare the results of multiple tasks, to receive hints or helpful feedback about cases where they were stuck or to coalesce their findings in some form of persistent inscriptions. The team could be encouraged and scaffolded to formulate summaries of their findings, noticings and conjectures. They could receive more systematic teacher feedback between sessions and have time to revisit previous topics armed with such feedback. Teams could share their findings in whole-class discussions after all the teams are finished with a given topic.

- Some cases require lengthy investigations and discussions, while others can be completed very quickly. Different teams bring different levels of mathematical experience and expertise to the curriculum. There should be a way for teams to pace their progress through the topics flexibly. That way, novice teams could spend more time enacting the basic practices in ways that are meaningful to them, while teams that are more expert (such as teams of math teachers) could move through the same set of topics faster and reflect on them at a higher level of mathematical sophistication.

- Although there was only a small set of instructions for the topic worked on in this team's final session, in general there are a number of instructions when a team works through several tabs of activity. It is always productive to revise the wording of the instructions based on the observed use of those instructions.

- In this last topic, it might have been more productive to have the team try to construct their own version of each quadrilateral right after they explored that figure. In that way, their observations would be fresh and could be immediately extended through the effort of re-construction.

What Did We Learn About What the Team Learned?

It is clear that the team learned a lot about collaboration, math discourse, GeoGebra tools and dynamic geometry. It is equally clear that they were still quite confused about the dynamic-geometry paradigm and the functioning of dependency relationships within dynamic-geometry figures. (This is not surprising given that dynamic geometry is a complex form of mathematics, which is not well understood even by adults schooled in Euclidean geometry.)

A major benefit of the preceding analysis of the Cereal Team's work in the VMT environment is that we were able to see the learning take place and observe the students' displays of understanding. This is rather rare in educational research.

Not only did we observe indications of what was learned, but we often could tell how it was learned—in general through the team's adoption of specific group practices. The lists of group practices we compiled are quite general and are probably applicable to a wide range of virtual math teams. Of course, each team's work is highly unique and situated in terms of the participants and the concrete flow of their discourse. However, most of the practices listed are typical components of learning about collaborative dynamic geometry.

In addition, we could judge quite well what worked in the curriculum of topics, with their example figures and their text boxes of instructions. So we can systematically revise the curriculum for the next iteration of trials with teams of teachers and of students. We can edit the detailed instructions, but also revise the selection, sequencing and timing of the topics. An example next version, incorporating many of the points listed at the end of the chapters on the eight sessions, is now available as (Stahl, 2015).

A central focus of the curriculum was on the notion of dependencies—how the design of dependency relations into the construction of figures preserves desired invariances. Rather than trying to provide broad coverage of geometry topics, the curriculum tried to facilitate a deep experience of this core theme of dependencies. It is well known that dependencies are hard for students to understand. However, dependencies seem to lie at the core of dynamic geometry. Perhaps one can say that they always underlay Euclidean geometry, but that dynamic geometry brought it to the more visible surface. We can see that the Cereal Team experienced some important insights into dependencies (especially in Session 6 and with Poly2 in Session 8). However, they also displayed serious misconceptions right up to the last session. The analysis of these experiences suggests ways to more tightly focus the curriculum on dependencies as an intermediate abstraction for the mathematical subject.

The CSCL pedagogy seemed to generally work well. Intervention of teachers as sources of authoritative knowledge was minimized during the online sessions. Teachers and curriculum developers stayed in the background, setting up the group sessions, providing the software infrastructure, drafting the topics with their instructions and offering motivation and feedback in classes before or after the sessions. We saw in many instances how the instructions guided the student work and discourse—although the student team always had to enact the instructions within the team interaction.

Finally, we even saw how the collaborative learning of mathematics mediates between individual and community knowledge. By seeing the differences between the three students and how those differences interacted and became shared, we had a sense of the zones of proximal development playing a central role. However, what we observed went past Vygotsky's sense of individual development being led by an adult or more developed peer. In the teamwork within the VMT environment, the interplay between the three zones resulted in group development that was more than the sum of its parts. The guidance was not directly from an authority figure, but was the referred intentionality embedded in the curriculum. The curriculum included resources going back to Euclid, as well as ideas and editing by numerous experienced educators and teachers. In addition, the GeoGebra software provided interactive mathematical guidance that is not available from paper-and-pencil drawings.

In the next chapter, we consider broad-level implications of the preceding fine-grained analysis of the Cereal Team's sessions for the theory of mathematical group cognition.

Contributions to a Theory of Mathematical Group Cognition

The bulk of this book consisted of the detailed review of all of the Cereal Team's sessions. From this, we saw that the Cereal Team quickly became an extremely collaborative team—as displayed in their productive mathematical discourse—and that they gradually developed a rich, but fragile sense of dependencies in dynamic geometry. We will now reflect on what we have observed of how they developed through the sessions as a collaborative team and how their discourse, their fluency with GeoGebra tools and their understanding of dependency grew over time through their work on the sequence of topics.

Having gone through the detailed analysis, we will here first summarize the process that emerged for the Cereal Team's development. In particular, we will try to further explicate our central research question: *How* did the team's mathematical group cognition develop?

Following the discussion of the developmental process, we will reflect more theoretically upon the six dimensions that guided the analysis:

(i) Collaboration and the development of group agency.

(ii) The discourse of mathematical dependency.

(iii) Dynamic-geometry tools mediating cognitive development.

(a) Dragging as embodied cognition.

(b) Constructing as situated cognition.

(c) Designing as conceptualizing dependency.

For each of these dimensions, we will elaborate a conceptual contribution to the theory of mathematical group cognition: group agency, dependency discourse, tool mediation, embodiment, situatedness and designing.

How Mathematical Group Cognition Developed

Although at different times during the sessions one student expresses more clearly than another an action to be taken—such as a construction step, a statement of a finding or an accomplishment—the three students work very closely together. They build on each other's actions and statements to accomplish more than it seems any one of the students could on her own. They often agree in the chat on each step and each conclusion. For instance, in each phase of the dual session (Sessions 5 and 6) involving inscribed triangles and squares in Topic 5—the explorative dragging, the experimental constructing and the determination of dependencies—the results are accomplishments of the group as a whole, as documented in our fine-grained sequential interaction analysis.

Of course, the topic instructions provide important resources and guidance in pursuing these steps. In addition, the teacher sometimes preps the team in class, before they meet online after school. However, once the team starts in the direction prompted, they do not simply follow the instructions. They become engrossed in teamwork that continues in a natural and self-motivated way, driven largely by the elicitation-and-response sequentiality of the team discourse. The relatively minimal instructions serve as more-or-less successful catalysts. They are necessary to guide the participants during early collaborative-learning experiences. In the future, the groups should be able to proceed when such "scaffolds" have been removed—as the students do in successfully constructing a square on their own without any specification of steps to follow. Furthermore, in the future, the individual group members may be able to do similar work by themselves (even in their heads) as a residual effect of their group work (in the chat and geometry software).

Throughout our analysis of the Cereal Team's sessions, we have identified lists of *group practices* adopted by the team. It is these group practices that largely account for the group's teamwork, for their ability to construct knowledge as a group and to think as a group. The group practices are also the mechanisms through which the individuals exchange and appropriate each other's perspectives and skills. The group practices are the keys to the development of group mathematical cognition, because it is through the adoption of these practices that the group develops. The answer to the question of how the group developed is that it successively adopted these various practices and incorporated them in its on-going interaction.

The team developed into an effective unit by adopting group collaboration practices. It developed its collective skill at working on dynamic-geometry tasks by adopting as a group diverse practices of dragging, constructing and tool usage. And it developed its ability to engage in mathematical problem solving by adopting dependency-related practices and practices of mathematical discourse and action.

The adoption process often followed a general pattern. First, the group encountered a "breakdown" situation in which they did not know what to do. Then someone made a proposal for action. There may have been a series of proposals, some ignored or failed and others rejected by the group. This may have been followed by a negotiation process, as group members questioned, refined or amended the original proposal through secondary proposals. Finally, there was often an explicit round of agreement. Perhaps most importantly, the new practice was put to work in overcoming the breakdown situation. In future cases, the practice may have been simply applied without discussion. Of course, there could also be instances of back-sliding, in which the group failed to apply a previously adopted practice where it could have helped. It should be noted that this general pattern is not a rational model of mental decision making. It is philosophically related to the theory of tacit knowledge, in which a breakdown leads to explicit knowledge, followed by negotiation and eventually a return to tacit practices (Heidegger, 1927; Koschmann et al., 1998; Polanyi, 1966; Schön, 1983; Stahl, 1993; 2000; 2006; Tee & Karney, 2010). The adoption process is driven by interpersonal interaction engaged in the world, not by logical deductions of an individual mind.

When we looked closely at examples of adoption of new practices by the Cereal Team, we saw their evolving use of terminology in chat and their increasingly confident use of tools in GeoGebra for joint exploration and construction. These analyses illustrate in detail additional ways in which the team adopted new practices.

The catalog of practices compiled in this book agrees well with lists of group practices enumerated in other studies. In a recent review of papers on VMT (Stahl, 2016b), many of the same practices were highlighted in the work of other student teams. These practices are by no means specific to group interaction in the VMT environment, but have been analyzed in numerous studies of CSCL and of interaction analysis (see e.g., Sawyer, 2014). More generally, Conversation Analysis has extensively studied: sequential organization (response structure), turn taking, repair, opening and closing topics, indexicality, deixis, linguistic reference and recipient design. CSCL has investigated: joint problem spaces, shared understanding, persistent co-attention, representational practices, longer sequences and the role of questioning. Within CSCL, studies of mathematics education have investigated: mathematical discourse and technical terminology, pivotal moments in problem solving and the integration of visual/graphical reasoning, numeric/symbolic expression and narrative.

The group practices identified here and many others work together to provide student teams with their unique and changing versions of mathematical group cognition. In the following sections, we touch on some general characteristics of such group cognition, related to the six dimensions that guided our analysis in this book.

(i) Collaboration and the Development of Group Agency

A plausible way of thinking about the development of collaborative practices at the group unit of analysis is in terms of increasing *group agency*. The concept of "group agency" has been hinted at increasingly in CSCL theory. It provides a helpful conceptualization for reflecting upon our analysis of effective collaboration in this book.

Agency is traditionally ascribed to individuals, indicating their ability to engage in intentional activities, in which they as subjects determine effects in the objective world. In particular, individual agency plays a central role in rationalist philosophies, where individual cognition and planning are assumed to underlie human action (for a critique of this view, see Suchman, 2007). Most references to "group agency" in academic literature are from an organizational-management perspective, rooted philosophically in a rationalism stemming from Rousseau (1762), in which a group or a society is conceived of in terms of an implicit contract among individual rational actors, each pursuing their own self-interest. Accordingly, for instance, the classic groupware systems for management focus on exchange of ideas among individuals (e.g., brainstorming) and decision making via voting mechanisms (see Stahl, 2006, esp. Ch. 7). The prominent recent book entitled *Group Agency* (List & Pettit, 2011) remains in this vein, focusing on contributions of individuals as independent agents.

An alternative approach to group agency has been indicated within CSCL by Charles and Shumar (2009) and Damsa (2014). This notion of group agency is consistent with the seminal sociological paper on agency by Emirbayer and Mische (1998). In its formulation—based on Heidegger's phenomenology of human temporality and influenced by Bourdieu, Giddens and Habermas—the authors propose that human agency be understood in a pragmatic, dialogical manner as a form of practice within the dimensions of experiential temporality. They characterize agency as:

> A temporally embedded process of social engagement, informed by the past (in its 'iterational' or habitual aspect) but also oriented toward the future (as a 'projective' capacity to imagine alternative possibilities) and toward the present (as a 'practical-evaluative' capacity to contextualize past habits and future projects within the contingencies of the moment). (p. 962)

This post-cognitive (Stahl, 2016b) definition of agency could be applied equally well to individuals or to groups. Sometimes, even large communities articulate their community agency, as in the American *Declaration of*

Independence ("We the people, ...") or Lenin's call to action (*What is to be Done?*).

We apply that view of agency to the interaction of a small group of students learning about dynamic geometry together. The discussion of the Cereal Team's interaction suggests a re-specification of the concept of agency at the small-group unit of analysis. That is, we can further remove the concept from the individualistic framing that the concept was subjected to in modern thought and free it to be grounded in social intercourse. By showing how agency can arise through small-group processes, we are able to see the basis of individual agency in dialogical negotiation and intersubjective temporality. Just as individual cognition is founded in group cognition and individual learning is founded in collaborative learning, so human agency is founded in group agency.

For many people, the major barrier to attributing agency to small groups of people in analogy to human individual agency is the fact that groups do not have a body. That is, there is no continuing physical substrate that underlies the existence of a group in the way that a person's body defines the continuity of that person's stream of consciousness, personality, goals and sequence of actions. For instance, Engeström (2008) bases his rejection of the term "groups" (in favor of ephemeral "knots" of people) in this observation, as do Schmidt and Bannon (1992) in their prohibition of "groups" from CSCW theory (for a rejoinder to them, see Stahl, 2011e). This is similar to the reaction against "group cognition" based on the fact that there is no persistent, material group brain.

However, Latour (2013) argues that the notion of a continuing substrate is an outdated and incoherent view, adopted by modern common sense from the tradition going back to Descartes (1633) and even Plato (340 BCE). Instead, Latour views all actors as sequences of non-continuous events, linked together by complex, open-ended networks of reference ("[ref]") and repetition ("[rep]"):

> As we have done up to now, we are again going to plant our own little signposts along these major trails to mark the branching point whose importance we have just measured. Let us thus use [rep], for reproduction (stressing the "re" of re-production), as the name for the mode of existence through which any entity whatsoever crosses through the hiatus of its repetition, thus defining from stage to stage a particular trajectory, with the whole obeying particularly demanding felicity conditions: to be or no longer to be! Next—no surprise—let us note [ref], for reference, the establishment of chains defined by the hiatus between two forms of different natures and whose felicity condition consists in the discovery of a constant that is

maintained across these successive abysses, tracing a different form of trajectory that makes it possible to make remote beings accessible by paving the trajectory with the two-way movement of immutable mobiles. (Latour, 2013, p.91f)

In Latour's recent analysis of modes of existence, the continuity of an agent is not grounded in a persistent substance, but in the dynamic interplay of (a) temporal processes of continuous self-reproduction [rep] or re-generation with (b) relationships of reference [ref] to an open-ended range of other actors.

Let us view the group existence (with its group cognition and group agency) of the Cereal Team in these terms:

- [rep]: The Cereal Team regenerates itself as a unified collectivity repeatedly through its discourse. At the start of most sessions, the team members greet each other and discuss what "we," as a team, should do. At the end of most sessions, they say good-bye, effectively suspending the existence of the team temporarily until next meeting. These conversational moves, member methods or group collaboration practices open and close each session of the group's repetitive trajectory of punctuated existence (Schegloff & Sacks, 1973). The elicitation-and-response structure of their discourse and of their GeoGebra actions (which are done in response to requests from others and as displays for others) during the session weave together to constitute group-agentic processes of decision making and decision implementation.

- [ref]: In addition, the team as a whole orients itself to the current topic and to its constantly shifting, complex situation through its verbal and geometric references to available interactional resources. For instance, in its rich discussion of Poly2 in Session 8, the Cereal Team references its past experiences, which involved constructing dependencies using circles. In addition to bridging back to their shared past and making it relevant to their present attempt to achieve a future goal, these references orient the team to the ancient community practices of Greek geometry. Interestingly, these group-memory references [ref] by the group to its own past (or future, or present) [rep] highlight both the intertwining and the non-linear characters of the [rep] and [ref] trajectories. This is connected with the lived-temporality structure articulated by Emirbayer and Mische (1998). Ever since Euclid, the geometry community has used [rep] the construction of circles to establish relationships of dependency [ref] within geometric figures. While the team works as a collective agent to use these references to pursue team goals, one can also analyze how the *individual*–level discourse utterances and GeoGebra actions of the students are influenced by this *group*-level mediation of *community*-level practices.

Since Latour's modes of existence are intended as a universal process ontology, they can be applied to every level of analysis, such as the individual, small-group and community. Each level can be analyzed as an intertwining of temporal streams of [rep] and [ref]. Their paths overlap in repetitions, as individuals interact in the groups and communities. They also reference each other continuously, as groups discuss the contributions of individuals and the resources, conceptualizations, social practices and artifacts of the community. Just as the group must constantly renew its unity and continuity through its temporal-response structure, so an individual mind must repeatedly build a stream of consciousness that references past events of its own and future goals. Similarly, a community must constantly replenish and refresh its symbols of identity, history, boundaries, ideals and vision. The individual mind can be conceived as a mini-group silent discourse with itself (as Vygotsky understood mental thinking to be a variant of silent talk); the VMT team as a discourse through chat and GeoGebra actions; the community as a super-group constituted by its common language (Sfard, 2008b). While inseparable in practice, the levels are often useful analytic conveniences or simplifications. They are conventions for identifying [ref] and conceptualizing patterns [rep] in the incessant buzz of [rep] and [ref] at different scales of the team interaction. There are also objective reifications (including Latour's immobile mobiles) of continuity at the different levels: individual personalities, group joint problem spaces or shared workspaces and prototypical constructions or whole texts like Euclid's *Elements*.

Perhaps the most striking aspect of the continuity of the agentic group is its ability to engage in extended sequences of discourse and action in order to succeed in solving challenging problems. As previously also documented in an analysis of how another VMT group structured its problem solving in SpringFest 2006 (Stahl, 2011a), the group process can sustain a discussion over time in order to integrate many contributions from individual members into a longer sequence of inquiry than any of the participants could sustain on their own. Skill in geometry requires the ability to project and carry out involved sequences of argumentation, analysis and construction. Some of Euclid's proofs include dozens of steps. While each step may be easy for even a naïve student to readily accept, the ability to design such a proof requires mature cognitive skills and sophisticated agency. For instance, in a famous dialog predating Euclid, Socrates leads an untutored slave through the multiple steps of a geometric proof. Even the slave, Meno, can agree to the truth of each step, but only Socrates can lead Meno through the proof trajectory from what is given to what is proven (Plato, 350 BCE). In the VMT context, the group agency of the Cereal Team leads the group through complex inquiries into the guiding topics, thereby providing experiences that help to develop the team members' individual agency, geometric skills and mathematical practices.

The development of group collaboration practices by the Cereal Team corresponds to an increase in group agency. As the students adopt the practice of taking turns in following the steps of the instructions, they in effect take turns in leading the group in its work. At first, the three students each try to lead in different ways: Cheerios repeatedly asks other students to take on tasks (without always seeming to understand what is already happening); Fruitloops poses challenging questions (steering the conversation in indirect ways); and Cornflakes experiments with GeoGebra tools (quietly investigating on her own technical opportunities for potential group exploration). As they establish collaborative practices, a shared group agency emerges, which incorporates and integrates their personal forms of agency. Thereby, each student becomes familiar with the other students' methods by responding to them and experiencing the results. Gradually, each also adopts or adapts the others' approaches, establishing them as shared practices. These group practices are also instantiated as individual practices within the group context. They will presumably be available for the individuals to use in other contexts in the future. As Vygotsky proposed with his analysis of the zone of proximal development, the *development* of group agency mediates the potential for the *learning* of individual agency.

(ii) The Group Discourse of Mathematical Dependency

The discussion of Poly2 in Session 8 provides a perspicuous example of the Cereal Team's productive discourse near the end of their developmental trajectory. Here, they discuss dependencies and they draw upon their previous experiences in order to further their work on their current topic. The concept of dependency has provided a central organizing theme in the VMT Project and in the current analysis. In Session 8, the team discusses this topic explicitly.

The Session 8 log from lines 28 to 91 is particularly complex to analyze, as we saw in the last chapter. There are overlaps in the typing of chat postings, leading to multiple threads of discourse. In particular, interchanges about the meaning of terms like "constrained" and "dependent" are mixed with explorations of the constraints among geometric objects in the GeoGebra tab.

To help sort this out and provide an overview of this interaction, the excerpt about Poly2 will be diagrammed below in a way that has sometimes been used for such discourse (e.g., Sfard & Kieran, 2001; Stahl, 2009b, Ch. 26). This excerpt was already discussed systematically in the section on Session 8 in which the students discuss terms of dependency as related to the use of the GeoGebra compass tool.

The group memory of using circles in constructions is central to the collaborative learning of the Cereal Team in this episode. The use of circles is the paradigmatic experience of dependency for the team. It is the only mechanism for establishing dependency that they have really discussed. They do not well understand the dependency associated with geometric transformations in Session 7, although they explored it briefly and might have developed a sense of it given more sessions and a better designed curriculum of tasks. They do not consider the simple case of dependency from Session 1, in which an intersection point of two lines is dependent on the lines, as a dependency mechanism.

In the beginning of Book I of Euclid's (300 BCE) *Elements*, the circle is used extensively (almost exclusively) to construct dependencies, which are then proven based on the equality of the radii of the circles. Since the initial topics of WinterFest 2013 were based largely on Euclid's presentation there, the predominant method of constructing dependencies in the VMT topics is with circles. In GeoGebra, this approach to defining dependencies with circles can often be streamlined by the use of the compass tool. The compass tool is essentially a custom tool that encapsulates the dozen construction steps of Euclid's second proposition (see Stahl, 2013c, Ch. 5.3). However, the Cereal Team did not differentiate between use of the circle tool and the compass tool in GeoGebra, which is an important but subtle distinction. Consequently, the construction of dependencies was primarily associated with circles for the team.

In several sessions, the Cereal Team engaged in shared attention to the visual display of constructions with circles. The team itself used the compass and circle tools of GeoGebra to construct equilateral triangles, perpendicular lines and even the rather challenging inscribed-square figure. Each team member went through the steps of manipulating GeoGebra tools and visible points and lines in the VMT tab to achieve these constructions. They each also watched as their teammates conducted the constructions. The team discussed the steps involved and the justifications for them, while visually observing the figures and physically manipulating them.

Sfard (1994) and others have argued that mathematicians develop "deep understanding" of abstract concepts with the aid of mental objects that are reifications, often linked to imagined visualizations as well as physical sketches. Our research question can be formulated as: How can young students who lack the conceptual skills of professional mathematicians or even of successful advanced math students begin to develop such skills? Our hypothesis is that this can be facilitated by involving students in experiences in which they are carefully and systematically guided to discuss mathematical issues with peers while sharing and physically manipulating appropriate visual representations. This could lead to the development and sharing of valuable discourse, visualization and manipulation skills. The synthesis of physical, verbal and social involvement could result through processes of reification, internalization, individualization, etc. providing something like deep understanding (involving multiple representations) to the individuals involved. Our analysis of the interactions of the Cereal Team were intended to investigate the nature and extent of changes in the implicated skill levels within the team interaction during its eight sessions of WinterFest—primarily by identifying group practices the team adopted.

Sfard (2008b) sees the development of mathematical cognition in terms of discourses, both in the historical process of the field of mathematics and in the individual process of learning mathematics. She notes that our discourses in math are historical repositories of complexity, which underlie our ability to build on achievements of previous generations rather than our having to begin anew every time. Husserl (1936/1989) makes the same point.

Sfard (personal communication June, 2013) argues that the idea of multiple levels of cognition—that of the individual, of a group and of community—becomes particularly credible if one takes discourse (e.g., geometric discourse) as the unit of analysis. The notion of discourse as the thing that changes when people learn geometry grows from the conceptualization of thinking as self-communication (as in Vygotsky). Once conceived in this way, thinking and communicating within a small team or within a community are but different manifestations of the same type of activity—they all belong to the same ontological category. They use the same artifacts (meaningful words of a community's language) and linguistic practices. This is quite unlike the situation that arises when one thinks about thinking as a unique type of mental process, distinct from what happens in interpersonal communication (speech) or culture (traditions and artifacts). The idea of considering all three levels is particularly convincing, indeed self-evident, when cognition is conceptualized in discursive terms. Equating thinking with a form of communication also implies that all three levels of cognition can be analyzed with the help of similar techniques of discourse analysis. For instance, one can analyze van Hiele levels of geometric thinking as a hierarchy of geometric discourses, in which each layer is incommensurable with the preceding one, but also subsumes and extends the adapted version of this former discourse. Individuals, teams and communities can move up the levels by adopting different discourses.

While we have been influenced by Sfard's focus on thinking as communicating (Sfard, 2002; 2008b) and on her methods of analyzing student discourse, we have come to some different results from, for instance, her example of Gur and Ari (Sfard, 2002; Sfard & Kieran, 2001). We use the term "interaction" rather than "communication" in order to emphasize that in the VMT context manipulation of GeoGebra objects is as much a part of the team interaction as the verbal chat postings. While the term "communication" can carry the traditional implications of personal understandings or mental intentions of the participating individuals (as in Shannon & Weaver, 1949), our concept of "interaction" is defined in terms of what is *shared by the group*. We try to capture in our data everything that is shared by the group (its whole history, its chat, its GeoGebra actions, and anything else which is visible to all the participants in the VMT environment).

We prefer to analyze small groups of three or four students rather than dyads, because our experience has been that dyads tend to fall into cooperative (at best) relationships rather than collaborative teams. In dyadic interchanges, it is too easy for both participants and analysts to attribute ideas to one person or the other, and it is too easy for the participants to fall into patterns of one person solving the problem and teaching the other (Cobb, 1995). In groups of more than two (but not enough to cause confusion in discourse-response structure), there is more chance for group cognition, where the group solves the problem together, step by step, and the major ideas or trajectories of investigation grow out of the situation of the interaction without being attributable to the individuals. Of course, collaboration must be learned by each new group, which is why we endeavor to motivate and teach effective collaboration the same way we teach productive mathematical discourse, namely through guidance in the instructions. We also attempt to analyze growth in collaboration skills in parallel to that of math understanding.

We try to minimize interpretive speculation about student intentions behind their actions or chat postings. This may be easier in our case than in the case of Ari and Gur because the focus of attention of the Cereal Team students is on objects that are visible in the VMT interface, such as triangles and movements of dragged vertices, rather than on mental abstractions like linear functions and their parameters. We do not just assume that Cheerios, Cornflakes and Fruitloops are generally watching the same visual display of their geometric objects, but we have evidence that they do that: as they take turns they generally imitate or build on each other's manipulations. References to labels of points and polygons facilitate a joint focus within the visual field. When someone creates points that are off someone else's screen (due to zooming or scrolling), there is usually a complaint registered in the chat. The affordance of a shared visual display—captured in the data for the analysts by the VMT replayer—

provides a basis for common ground (for the participants, but also for the analysts) that does not require the same kinds of assumptions or speculations as conceptual understandings (Stahl et al., 2011).

We can illustrate our approach with the confusing chat about Poly2, reproduced in **Log 63**.

Log 63. References in the chat about Poly2.

Line	Start Time	Post Time	cheerios	cornflakes	fruitloops	GeoGebra	References
28	19:55.5	19:58.3		ill go next?			regulate collaboration
29	20:12.5	20:13.2			sure		
30	20:16.6	20:26.5		...poly2 ofg?			poly2
31	20:31.6	20:37.5	justify...			Gino...akes drags E, F, G, H	point labels
32	21:00.4	21:17.3		...has			poly2
33	21:17.6	21:28.8		...ant me all... ferent			point G
34	21:33.4	21:40.3	do u think it a...				"restricted"
35	21:41.3	21:44.7	or constrained				"constrained"
36	20:51.0	21:49.4			...ke poly 1 and po..		color of points
37	21:59.6	22:04.5			...moving n3		regulate collaboration
38	22:16.3	22:17.1		sure			
39	22:16.1	22:25.0			...cheerios, I dont know		
40	23:09.9	23:18.0	ok can..			Fruitloops drags G	regulate collaboration
41	23:21.2	23:22.7			...yes		
42	23:07.5	23:23.3			...going only m..ves in..		point G
43	23:31.3	23:35.6		..ruitloops yea			
44	24:01.0	24:16.7	what ..he d..ference				"restricted," "constrained"
45	24:22.1	24:24.3	n*				repair
46	24:35.0	24:41.6		c..strained is limited		Fruitloops drags E	"constrained"
47	24:23.3	24:46.4			..when you m..e..		points E, F, G
48	24:54.2	25:08.4			so I think e it fi..init..		point G "constrained"
49	25:13.2	25:14.0			yea..		
50	25:16.1	25:19.8			..too		
51	25:23.4	25:25.6	why though				
52	25:36.5	25:59.3			and g moves whenever y..		points E, F, G, H
53	26:17.7	26:20.3	okay				
54	26:07.0	26:42.4			..i think its		"constrained"
55	27:33.3	27:36.8	oh i see				
56	27:28.4	27:37.5			..is the definition of		"dependent"
57	28:38.0	28:52.4	u need..g other...				construction question in
58	28:43.6	28:54.6			do you guys h..any id..?		"dependent"
59	28:57.1	29:15.6					dependent shapes
60	29:23.4	29:43.4		maybe..me in..ti..ive			equilateral triangles
61	29:49.7	30:02.4	yea maybe like th..				points F, G
62	29:51.3	30:20.3			..because point g o..		points F, G
63	30:32.7	30:35.3		but w..			
64	30:45.5	30:55.5			..a ba.. ..with h..		construction question in
65	31:02.4	31:03.8	i agree				
66	31:25.1	31:29.1		its			
67	30:57.4	31:44.4			cause erem..er how		dependent shapes
68	31:50.4	31:58.9		..of the p..t			
69	32:17.3	32:38.1	yeah th.. ma.. ..se				equilateral triangles
70	32:42.3	32:43.0		yes			
71	33:00.4	33:10.4			but..n't really kno..how it		construction question in
72	34:03.8	34:14.5	maybe they..ed a..her..				dependent shapes
73	33:59.6	34:17.4			..think point d is the		points E, F, G
74	34:13.7	35:03.6		we po..ts ha..			dependent shapes
75	35:06.1	35:31.9	i think it is the sa..				compass tool
76	36:02.3	36:13.9			an.. h is just co..letely	Fruitloops stops dragging	point H
77	36:21.3	36:30.8			yeah a..eeb.. ..		construction question in
78	36:26.9	36:31.7	agreed				
79	36:36.2	36:37.6		no..			
80	36:38.5	36:55.4			..a was probably th.. f..		construction question in
81	37:03.1	37:05.9	yeah				
82	37:11.5	37:14.9		..in..			
83	37:25.2	37:33.2	the first p.. ..c..				construction question in
84	37:41.7	37:51.7		..in a ..h..etical			
85	38:02.4	38:06.7		e..and then h			
86	37:21.4	38:16.7			..ere..ve..ywh..re		color of points
87	38:28.9	38:45.0		..e and f are the a..			color of points
88	38:51.1	39:02.3			..e and f are constrai..		"constrained"
89	39:06.0	39:14.6			..nt know for sure may..		
90	39:51.6	39:57.6	im not very sure e..			Gino...akes drags E, F, G, H	
91	40:00.5	40:02.4		..neither			

In this representation of the chat, the times are given for when each chat posting started to be typed and when it was actually posted, so that the overlaps of typing can be taken into account. The postings of each student are displayed in a different column, to provide a visual impression of the flow of interaction. In addition, arrows connect posts as follows: Black arrows indicate where one student builds on her own previous posts [individual rep]. Green arrows indicate where one student elicits a response from others. Blue arrows indicate where one student responds interactively to a post by a different student (analogous to what are called adjacency pairs in Conversation Analysis—the glue of [group rep]). We have included prospective elicitations of response as well as retrospective responses, in accordance with Sfard's method (Sfard & Kieran, 2001). Note that every elicitation is responded to—which in itself could be considered an indication of effective collaboration.

Rather than speculating on focus or intention as in Sfard's analysis, we take into account on-going GeoGebra actions and mathematical objects referenced in the chat, which provide the generally shared focus [group ref] for the group activity. One column lists associated or simultaneous GeoGebra actions. A final column lists some of the objects or terms that are referenced in a given post [individual ref]. These references may be to geometric objects in the GeoGebra tab, to special terms that the students have used in their discourse, to words and phrases that appeared in the instructions in a GeoGebra tab, to geometry content [community ref] or to previous experiences that the team has shared in prior sessions [group ref].

In scanning the flow of interaction visualized in the spreadsheet with arrows, one can notice the important role of questions as frequently providing "pivotal moments" in the thematic content and threading of responses (Stahl et al., 2011; Wee & Looi, 2009; Zhou, 2009). This is because the pragmatic role of a question is generally to elicit a response. The response may form a major adjacency pair with subsidiary sub-pairs (Schegloff, 2007), thus potentially forming a "longer sequence" (Stahl, 2011a). The [rep] and [ref] trajectories of individual, group and community modes of existence are not objectively given, but are constructed interactionally and interpreted narratively.

Learning how to initiate and contribute to longer discourse sequences is an integral component of learning to formulate mathematical arguments (explanations, justifications, proofs, axiomatic deductions). Mathematical arguments are instances of longer sequences. The ability to participate in and eventually initiate such longer sequences is an interactional and eventually a cognitive skill that has to be learned and developed, initially through discourse with others. This is a social and cognitive skill or practice necessary for mathematical thinking, similar to the ability to construct narratives, so central to oral society (Bruner, 1990; Ong, 1998; Orr, 1990). That is why it is so significant that the practice of questioning (e.g., "why though?") is gradually shared by the three team members. In the early sessions, questioning is characteristic of Fruitloops' contributions to the discourse. In the later sessions, Cornflakes and then Cheerios adopt this role. The questioning is often prompted by the wording of the instructions visible in the GeoGebra tabs.

We have circled a number of postings in yellow to indicate that we have identified them as potential "pivotal moments" (Wee & Looi, 2009). By this final session, the team has developed several group practices related to posting potential pivotal moments. For instance, in this excerpt concerning Poly2, it is striking how the three students repeat the use of the term "maybe" in lines 60, 61, 62 and 72. This term functions to introduce a tentative conjecture. It secondarily serves to tie these postings together in an extended sequence of intense speculation about the construction of the dependencies in Poly2. The students seem to have adopted the group practice of marking their conjectures as tentative. In addition to introducing them with "maybe," they use the hedges "probably," "could have" and "I think" in this excerpt.

Another practice which introduces potential pivotal moments is questioning "why though" or "but why." Fruitloops used these locutions in the early sessions, but the others use them in the later sessions, including in this excerpt. A more common potential pivotal move is to use standard English interrogative forms, such as "what is" or "do you think." It is interesting that of the 15 postings identified with yellow circles in the figure of the excerpt, each student contributed the same number, five. This shows a striking equality of initiative in contributing to the group agency. Note that the pivotal postings tend to form particularly dense foci of elicitation and response arrows, and often initiate "longer sequences" of threaded interaction.

The density of arrows in the figure indicates a healthy level of interaction, with most of the elicitation moves being met by responses, a distribution of initiative and an attention to what each other is saying. Comparing the approach taken by Ari and Gur to that by Fruitloops, Cornflakes and Cheerios, we see advantages which might be attributable to the VMT approach. These include: (a) the guiding wording of the VMT instructions, (b) the attempt to provide training in collaborative interaction, (c) the presence of a visual representation providing a shared, persistent "group memory" (Çakir et al., 2009), "common ground" (Clark & Brennan, 1991) or "joint problem space" (Roschelle & Teasley, 1995; Sarmiento & Stahl, 2008a), (d) the team size larger than a dyad and (e) the longitudinal analysis as opposed to two snapshot excerpts.

All these factors may have contributed to the contrasting outcomes. In the case reported by Sfard, Ari always solved the math problems individually and then shared the result with Gur, but without much concern for Gur's understanding of the solution process. In the case of the Cereal Team, solutions were often obtained that none of the three team members could have reached on their own, and the successive steps of the solution were contributed by different members, resulting in an achievement of group cognition. The Cereal Team made sure that each member tried out the solution in GeoGebra and agreed with it before moving on. Team discourse picked up on technical terms from the instructions, which gradually became better understood as different people applied them to prototypical cases.

While analysis of an early and late excerpt—if strategically selected—can indicate progress or lack of it during the intervening interval—much like a pre- and post-test—it is not likely to display *how* the progress was achieved or prevented. In contrast to Sfard's analysis of two excerpts of Ari and Gur's communication, our analysis looks at the entire continuum of interaction within the Cereal Team during its eight-session existence. While Sfard was able to conclude that the communication style of Ari and Gur contributed little to their mathematical learning, our analysis tried to document how the interaction of our team evolved and to provide evidence for explaining how particular interaction events contributed to improvement of their collaboration, their discourse and their mathematics. We have attempted to identify displays of learning in our data, and to associate these with details of the design of the VMT environment in order to derive implications for re-design of our educational intervention. Our analysis aimed at highlighting features of our experimental environment (technology, curriculum, guidance, organization) that promoted or hindered progress in the team's and the individual students' improvement in collaboration practices, in productive discourse, in use of GeoGebra tools and in understanding of dependency in dynamic geometry through dragging, construction and designing dependencies.

We have remarked on a number of features of the Cereal Team's *productive mathematical discourse* that may have contributed to their increased ability to engage in longer sequences and to spend more time-on-task discussing mathematics. Their mirroring of each other's discourse moves—like the use of "maybe" hedges or "why though" pivotal questions—serve both to align the group members in engaging in interaction and to elicit continuing responses. The topic presentations—initially structured with numbered steps—led the group through sequences of tasks and prompted for associated discussions, providing thematically coherent stretches of interaction. The sharing of each other's personal approaches increased the portfolio of moves available to each and helped them to understand each other's actions from their own perspective and experience.

(iii) Dynamic-Geometry Tools Mediating Group-Cognitive Development

The VMT Project is premised on the hypothesis that engaging groups of students in carefully designed and supported dynamic-geometry activities can foster their development of mathematical group cognition. To explore this hypothesis, GeoGebra, a specific implementation of dynamic geometry, was integrated into the VMT collaboration environment. We then looked at how the student group enacted the tools of collaborative GeoGebra. The concept of "instrumental genesis" (Damsa, 2014; Lonchamp, 2012; Overdijk et al., 2014; Rabardel & Beguin, 2005; Rabardel & Bourmaud, 2003; Ritella & Hakkarainen, 2012) may be helpful for conceptualizing the way the Cereal Team gradually increased its mastery of GeoGebra tools and practices, such as dragging with the move tool, constructing with the compass tool or creating patterns with transformation tools.

The central lesson of the theory of instrumental genesis is that tools are not simply "given" for people or groups learning to use them. The nature of the tool must be "enacted" in the use setting by the users (LeBaron, 2002). The theory of instrumental genesis is part of a larger post-cognitive philosophy, which rejects the realism of the rationalist tradition that culminated in cognitivism. Post-cognitive philosophy (Stahl, 2016b) avoids the charge of relativism by grounding the enactment of reality in a dialectic of "creative discovery." While capable of being enacted in an open-ended variety of ways, the characteristics of the created reality are discovered in a "reflective conversation with the materials" (Schön, 1983). The Cereal Team tries out proposed conjectures about reality by dragging existing GeoGebra objects and trying to construct new ones. Their views of the GeoGebra micro-world are delimited by their explorations and experiences with its objects. Individuals are not free to construe reality arbitrarily, but are constrained by the social, embodied and situated results of enactment efforts.

Kant (1787/1999) argued that the human mind constitutes meaningful reality through a process of creative discovery, in which structure is imposed by the mind on reality in order to create-and-discover objects in the world. In the preceding analysis of the Cereal Team, we see how group interaction can constitute the character of objects in the shared world and how the shared meaningful world is itself constituted through such interaction. The nature of reality—such as the dependencies of inscribed squares—is discovered through the creation of interpretive views of objects. Effective perspectives are constrained by reality, which is not knowable except through these views and interventions. The creation of perspectives at the level of group cognition shifts the constitutive role from Kant's individual cognition to group and social cognition (Stahl, 2016b).

Students in virtual math teams learn to see things *as* others see them in group-cognitive processes (which generally incorporate culturally sanctioned community approaches). Subsequently—due to the power of language (e.g., naming, verbal description, articulated remembrances)—the students (individually or as a group) can "be there" with those objects (squares, segments between vertices, dependencies)—even when they are not physically (or digitally) present with them—in a shared group setting. Although not visibly present, objects can "be there" in imagination, in the recalled past or in a projected future. People can even "internalize" (to use Vygotsky's metaphor) their ability to be-there-with these meaningful objects in the internal speech of individual thought (imagining, remembering, projecting). The fact that introspection by adults discovers (and assumes) the existence of many individual mental objects does not mean that those objects were not at some point in our development internalized from group-cognitive experiences in community contexts. An adequate analysis of cognition should recognize the constitutive roles of group cognition and their integration with phenomena of individual and social cognition.

In particular, cognition—at all levels of analysis—is *mediated by the available technology* (Carreira et al., 2016). This is especially apparent in a context like the VMT environment, where tasks necessarily involve the use of special software tools. The team learns to construe the problems it is given in terms of the affordances of GeoGebra: e.g., dragging figures to explore their constraints and constructing figures with various tools to establish dependencies. The instructions associated with the VMT curriculum guide the team to approach its tasks through the use of appropriate GeoGebra tools. Thus, the team's understanding of the usage of GeoGebra tools is essential not only to solving the geometry tasks, but even to understanding what the problem is and how to approach it.

Note that in Session 3, involving the construction of perpendicular lines, the students had already had experience with the use of the relevant tools in their previous sessions. The curriculum had been designed to cover the tools and practices of perpendicularity and sequentiality before it challenged students to construct a perpendicular in Session 3 or a square in session 6. These are procedures that require specific sequences of construction actions, including several involving locating special tools and constructing accessory circles to establish dependencies. In both of these cases, it took the students considerable experimentation with the use of the involved GeoGebra tools to become skilled at using them as needed. In addition, it took a while for the team to even recognize that they should be applying tool-usage practices that they had begun to establish in previous sessions. The idea of taking advantage of previous practices is itself a group practice that had to be adopted. Whereas it took a long time in session 3 for the team to start applying past construction practices, in later sessions they did this much faster, eventually automatically.

In working on Topic 5, it took the team many minutes of complex exploration of the affordances of various GeoGebra tools—especially the circle and compass tools—to discover sequences of actions that would construct the inscribed triangle or the exterior square. This included construction and dragging actions as well as narrative acts to persuade the group of the construction's adequacy. In each case, the end procedure was a relatively long sequence. It was not possible for any one student to implement the whole sequence on her own at first, but the team as a whole could. The individuals could observe this as a practice adopted by the group and then could potentially eventually implement it on their own. For this achievement of mathematical group cognition to be possible, it was first necessary for the team to have established not only the elementary tool-usage, dragging and construction practices, but also the collaboration and discourse practices, such as persevering in sequentiality. The VMT curriculum had been designed to prepare the way by guiding the team to engage in these practices in earlier sessions.

In the following, we list some issues involved in the social, embodied, situated and constrained enactment of the GeoGebra dynamic-geometry application and its software tools—as they emerged in the interactions of the Cereal Team through collaborating, dragging objects, constructing figures and designing dependencies. We have already extensively discussed the social nature of group cognition: that the character of reality is creatively discovered as a shared world—shared by the group and its community—ultimately to varying degrees by the universal community of humanity.

(a) Dragging as Embodied Group Cognition

Dragging is the most prominent feature of dynamic-geometry systems. It is what makes this geometry dynamic. Typically, a static geometric figure only shows one possible position of a geometric object or configuration of objects. However, propositions about geometry generally apply to whole ranges of different positions or configurations, all of which correspond to or are consistent with the set of conditions specified. Dragging allows a figure to go through many of these possible positions.

Experienced mathematicians can imagine a figure changing its position and appearance, but novice students do not yet have this mental skill. For instance, Cheerios typically draws a triangle in the prototypical position of an equilateral triangle with a horizontal side on the bottom. She draws a perpendicular as a horizontal segment with a vertical segment rising from it. She discusses these as fixed shapes in the early sessions.

The dynamic-geometry software allows students to drag a figure into new positions and observe its changing shape. The action of dragging a point on the screen with a mouse or touch-pad gesture provides a visceral experience to the student's muscles and active body (Merleau-Ponty, 1945/2002). Seeing the point being dragged and watching the consequences this has for the figure provides a visual experience for everyone watching. This embodied cognition provides grounding for future imaginative varying of figures. As Lakoff and Núñez (2000) document, mathematical cognition requires such bodily grounding in order for people to make sense of it. Our concepts of shape and space are largely metaphorical extensions of our bodily orientations. We have seen in the sessions that the three students spend considerable time dragging objects around. At first, they are resistant and tentative. They are not used to moving geometric figures around or changing their visual shapes. Then they observe things when they are doing the dragging themselves. Later, they are able to make significant observations when a teammate is dragging.

There are many roles that dragging has to play in work on dynamic geometry. Here are some of the roles we observed in the work of the Cereal Team:

- To give a student a visceral sense of geometric motions and relations.
- To give a student a visual image of figures, variations of figures and fixed relationships, which are maintained and cannot be altered (given the constructed dependencies).
- To bring multiple geometric objects into relationship with each other.
- To modify the shape of a given geometric figure to see that some features remain and others change.
- To explore what points or figures can change position or shape and which are dependent and can only be moved indirectly.
- To test that a figure satisfies specified conditions as stated in a problem or question.
- To investigate ones conjectures about fixed features by testing whether they can be changed.
- To check that a construction maintains intended relationships by trying to violate them. (This is the "drag test.")

(b) Constructing as Situated Group Cognition

By constructing a figure, students in a VMT team create a context of on-going work, which is visible to the group. They can then conduct a discourse situated in that shared context. The evolving situation is observable by the group in the form of the VMT interface on their computer screens. This shared workspace affords shared co-attention and the visual elaboration of the group's joint problem space (Sarmiento & Stahl, 2008a; Teasley & Roschelle, 1993). It serves as a working group memory, displaying the group's recent chat comments, constructions and knowledge products (Çakir et al., 2009).

The GeoGebra construction tools and how they are used become part of a group's or a student's conceptualization of dynamic geometry. By working out ways of using the available tools, groups of students construct what Hoyles and Noss (1992) call *situated abstractions*. These are ways in which people make mathematical sense of the results of their actions. They are sense-making devices, which are situated in that they are derived from concrete experiences within specific mathematical situations. However, they are also abstractions in that they operate beyond the specific experiences in which they arise, as they become generalized as group practices—shared and accepted ways of using the tools. Concrete, particular figures visible and manipulable on the computer screen contribute to abstract concepts like dependency, adding to the complex of multiple routines (Sfard, 2008b) reified in the technical term.

Most of the Cereal Team's productive discourse was centered on constructions—either given example figures or the team's own constructions. Construction was central to their inquiry processes. Their construction action often served as communication actions, as they showed each other how to do things (Çakir & Stahl, 2013).

As Damsa (2014, p. 8) says, it is important to gain insight into how students work together to construct and develop knowledge products. We have tried in this book to document and analyze small-group learning to reveal in detail how knowledge objects (such as inscribed triangles) are literally constructed and how they emerge from the group interaction. We have seen how the Cereal Team has engaged in dynamic-geometry construction activities to accomplish the following:

- To give the students a visceral sense of building geometric figures.
- To give the students a visual image of figures being constructed and dependencies being imposed.
- To test ideas for figures.
- To test procedures for building figures.

- To test procedures for imposing dependencies or relationships.
- To test that plans are correct and complete.

(c) Designing as Group Conceptualizing of Dependency

In the beginning of this book, we claimed that gaining a sense of dependencies by designing figures in dynamic geometry could provide a watershed experience for students in the kind of thinking that is foundational for science, technology, engineering and mathematics (STEM). The VMT Project has taken the understanding of dependency as a knowledge object at the core of dynamic geometry. A full understanding of this concept is far beyond the reach of middle-school students and, in fact, has not been explored much even in the academic research literature surrounding dynamic-geometry education (Stahl, 2013c). More generally, the intimate connection between explanatory proofs in geometry and dependencies underlying invariants has rarely been noted in the literature, although it may be implicit in the developmental approach of van Hiele's theory—that one must understand relationships before one can construct formal proofs (at least "explanatory" proofs).

The development of mathematical cognition is a long process. The final stages of the van Hiele levels, for instance, involve formal deductive proofs and axiomatic systems, usually not mastered until advanced college courses. This is a highly abstract manner of thinking and speaking. As van Hiele (1999) recommended, it is important to lead students to such abstractions through experiences with concrete phenomena, such as the designing and manipulation of simple geometric figures.

Scardamalia and Bereiter (2014) discuss *knowledge building*, applying this concept at all levels: individual, small-group and community cognition. Knowledge building involves the ability to create and refine knowledge objects such as texts, explanations or designed figures. It is particularly concerned with knowledge in the form of "designs, theories, problem solutions, hypotheses, proofs and the like" (p. 397). In our analysis of the Cereal Team, we have observed the students begin to create such knowledge objects. As they explored, designed and created figures in GeoGebra, they refined the concepts used to reflect on the figures together. We have identified a large number of group practices that increased the team's ability to do both the hands-on and the verbal knowledge building. Other researchers have described similar knowledge-building practices at the individual and classroom levels.

The Cereal Team gradually refined its understanding of the concept of dependency through its interactions during its sequence of eight sessions. The team's verbal knowledge building was facilitated or mediated by its adoption of group collaboration practices. The team developed its hands-on knowledge-building capability through the adoption of group dragging practices, group construction practices and group tool-usage practices. Finally, the team's ability to engage in conceptual knowledge building around the notion of dependency was enabled by its use of group dependency-related practices and group mathematical-discourse-and-action practices. For purposes of analysis and presentation, we have distinguished these categories of knowledge building and of group practices; however, they are all integrated in the actual group interaction.

Having followed the developmental trajectory of an impressive team of students working through the curriculum of eight sessions in the VMT activity system, we may conclude that attaining a robust sense of dependencies was a difficult challenge for these young students, individually and as a group. While we observed Cheerios, for instance, noticing a key dependency in the inscribed-triangles example figure and also constructing the dependencies required to form a square, there were other times when she did not seem to understand dynamic dependencies very well at all. In earlier sessions, Cheerios talks in terms of static appearances or shapes. Even in the final session, she describes a particular quadrilateral as a "square"—despite having just dragged it into a crossed quadrilateral and despite the fact that the final static appearance is clearly not a square with right angles. She has still not fully adopted the established dynamic-geometry paradigm.

Given the only partial success in guiding the Cereal Team to understand dependency in dynamic geometry, it seems we need to further refine the curricular resources to focus even more explicitly on dependency. Some ideas for refinements are to:

- Provide several models of dependency (point on line, point at intersection, equal circle radii, transferred compass length, rigid transformations).
- Guide discussion of dependencies more closely.
- Guide observation of dependencies more closely.
- Guide dragging to discover dependencies.
- Guide dragging to confirm conjectures about dependencies.
- Guide drag test for testing dependencies, including special cases.
- Guide planning construction of dependencies.
- Guide explanations using dependencies.

- Illustrate use of dependencies in proof.

These refinements will be incorporated in future versions of the VMT curriculum (such as Stahl, 2015).

Constructing Dynamic Triangles Together

The Dialectic of Mathematical Cognition

Group cognition is not something "possessed" by the group, the way knowledge was once conceived of as a possession, acquisition or discovery of an individual mind. It is perhaps better thought of as a process in which individual group members interact, thereby influencing and enhancing each other's ability to act in the future. While one may be able to claim that individuals are doing everything, the interaction becomes so complex that it can only be followed as a process of the group as a whole. Actions of individuals—including cognitive acts—when effectively collaborating are so subtly emergent from the interactional context that they cannot simply be attributed to the individual, but must be considered products of the interactional context, which is primarily the life of the group—as mediated by the activity system of tools and community practices. As we have seen in interpreting the discourse of the Cereal Team, the meaning of their chat postings are defined interactionally by their sequential role in eliciting and responding to each other. The meaning of the words and postings are defined by their use in this interaction. The meaning of a posting is determined by how it is taken up by the group. Interpretation (by the participants or by observers) involves a hermeneutic dialectic of whole and part (discourse and its constituent words or utterances) (Gadamer, 1960/1988).

Designating group cognition as a *dialectic* is meant to incorporate several aspects. First, that the group and the individual constitute each other mutually through their interplay. The group is nothing more than the set of its individual members (as they interact within their many-leveled context). Conversely, the individual is a product of the group behavior, whose actions are responses to the group setting, its goals, its history and its momentary circumstances. In the VMT environment, the group only exists for a delimited time interval, whereas the individuals come to the group with a much longer past and an indefinitely long future (as part of various other groups, like their math class or their family). So, people tend to assign a priority to the existence of the individuals. However, in the analyses of this book, the group often exerts the dominant role in determining individual actions. The term dialectic also emphasizes the dynamic nature of the individual/group relationship and of the interactional process. The dialectic proceeds with its own logic, driven by tensions or contradictions in the interplay between the individuals and the evolving group.

Perhaps the ontological characteristic of mathematical group cognition hardest to comprehend is its probabilistic nature. What the group "knows" at any given time is not some fixed set of facts that could be expressed in propositions and that the group either knows or does not. Rather, it is a varying chance that the group will be able to respond in certain ways to different opportunities that might arise. For instance, when the Cereal Team successfully constructed the inscribed squares, they did so because the group had experienced a number of previous involvements, which put them in a position to solve the challenge, but certainly did not guarantee it. The team had learned to collaborate well, which allowed the team members to build on each other's work: articulating the insight, using the compass tool and dragging the polygon. They had developed productive mathematical discourse so that they could spell out a plan and all understand it. They had reached an adequate level of skill in using GeoGebra tools, such that at least one member could do each necessary construction or dragging task. In addition, they had understood—most immediately from the inscribed-triangles problem—how to explore, design and construct dependencies in dynamic-geometry figures well enough to locate the vertices of the inscribed square. However, none of these distributed group skills or practices were firmly enough established that they could be put into action automatically. It took the group considerable trial and error to solve the challenge. There was a good chance that they might not have succeeded. For each of the things the group had to do, there was only a certain (non-quantifiable) probability that they would in fact succeed in doing it.

The probabilistic conception of knowledge can be applied to individual capabilities as well, particularly within the Vygotskian zone of proximal development. For an individual to be within her zone is for her to be ready to be assisted into accomplishing some cognitive achievement that she probably would not yet be able to accomplish on her own. As she develops through interacting with others around this achievement, her probability of being able to accomplish it increases. We already saw in Session 1 that Fruitloops, Cornflakes and Cheerios seemed to have different initial strengths. Cornflakes did not start out knowing how to use GeoGebra tools, but she was more oriented toward trying them out than were the others. As her tool knowledge increased through trial and error, through the step-by-step guidance of the topic instructions and through suggestions or examples from her teammates, Cornflakes increased the probability that she could successfully construct figures called for by their work context. Each of the team members had different probabilities that they could achieve certain kinds of tasks: leading the group, raising relevant theoretical issues or constructing dynamic-geometry figures.

These probabilities changed over time as the individuals developed and shared their skills. This individual development was inseparable from the development of their team's mathematical group cognition—the probability that the team as a whole could collaboratively accomplish the tasks. We have used the expression "fragility of knowledge" to indicate that at various points during the eight sessions there was a probability that the team could accomplish a given task, but not a guarantee. This probabilistic nature of mathematical group cognition makes its analysis, assessment and designing complicated.

Implications for Designing

What design principles have we discovered?

A general approach to design of collaboration environments for promoting the development of mathematical group cognition—including curriculum design, pedagogical approach and technology affordances—would be to support the adoption of group practices such as those we have observed for collaboration, math discourse and tool usage.

Here are some specific design principles based largely on observations of where the Cereal Team ran into difficulties (breakdowns) in their interaction within the WinterFest 2013 environment:

1. Support synchronous discourse; support multi-user visualization and manipulation; support turn taking of construction; support persistency and review of history; support open-ended exploration.

2. Use carefully worded and structured instructions; step users through initial sessions and provide prompting to model collaborative practices; repeat and refine technical word usage and prompt for adoption.

3. Sequence guidance in the development of group practices so that those practices needed by or useful for the adoption of new practices are already established and are likely to be enacted by the time they are needed.

4. Scaffold tool usage and gradually remove detailed scaffolding; carefully sequence tasks to build tool usage; encourage users to take turns and to all try tool usage.

5. Provide paradigmatic examples of dependencies; explicitly point out dependencies; prompt for discussion of dependencies; provide model explanations; prompt for explanations using dependency relationships.

In the next iterations of courses in teacher professional development and in WinterFest events, we should simplify the technology to provide just the necessary supports and refine the instructions to provide clear guidance and a clean sequencing of topics covering techniques of dragging and constructing as well as paradigms of dependency (intersection, circles, compass, transformations).

These implications largely motivated the latest round of development in the VMT Project. Based on the analysis of the interactions and achievements of the Cereal Team and other student groups in WinterFest 2013, the VMT environment was extensively re-designed for WinterFest 2014 in the following ways:

- Of course, the collaboration software was further developed to eliminate known bugs and to introduce new features. However, the major change was to curricular resources. The teacher-professional-development course was focused more on construction of geometric dependencies, giving teams of teachers considerably more hands-on experience with the kinds of tasks that their students would face. The WinterFest curriculum was extended from 8 sessions to 10 sessions, but the number of tabs for each session was reduced to about half as many and the tasks were simplified so that teams could be expected to complete all the work within one-hour-long sessions.

- The WinterFest 2014 curricular resources are restricted to dragging and construction of basic dynamic-geometry objects and exploration of the characteristics of triangles. The use of the compass tool for defining dependencies is presented in detail and the construction of isosceles and equilateral triangles are explored extensively. YouTube videos are included, illustrating clearly the role of the "drag test" and the use of the compass tool. In addition, students are involved in programming their own construction tools, so that they understand more intimately how dependencies are constructed in dynamic-geometry systems.

- Students are given workbooks, which motivate the topics in the tabs, provide some background narrative and provide spaces for students to record their observations and questions (Stahl, 2014c). The wording of the instructions for each of the topics has been edited for clarity. The text now emphasizes the consideration of geometric dependencies. Students are encouraged to preview upcoming topics and to continue work that their team did not complete. Teams are encouraged to return to complete or reconsider work on previous topics.

- Because a variety of arrangements are organized for student groups to participate in WinterFest—such as after-school math clubs, in-class lessons, at-home networking—the teachers are given considerable latitude in how they facilitate the groups. However, the teachers have been involved in reflecting on their own group's work during the professional-development course and they are required to summarize the work of their student groups. They receive credit for their involvement in WinterFest and are prompted to set pedagogical goals for their students' involvement and to compare these goals with perceived achievements. Teachers often gather WinterFest participants together between sessions for feedback, discussion and reflection.

- Students are given increased experience with construction.
- Students are given more explicit exposure to multiple paradigms of dependency. They are given more hands-on construction of dependencies, starting with the isosceles triangle as a clear example of imposing a dependency.
- The topics have received on-going refinement of their wording, so students will pay more careful attention to the precise wording as hints to the mathematical meaning and to rigorous mathematical language.
- Students are led more systematically through transitions in their discourse/thinking from referring to visual shapes, to drawing figures, to measurement of lengths or angles, to use of mathematical symbols, to engagement in dynamic constructions.

Advances and Future Prospects

When the VMT Project started, its goal was simply to create an environment in which students could talk about math in small groups online. Twelve years later, the VMT research team meets weekly to look at sessions of students talking about math in interesting ways. In particular, we looked during the winter of 2013/2014 and later at the discussions that three middle school girls had about dynamic geometry. Their discussions strongly evoke a sense that they are having the kinds of discussions that a few Greeks had in the 5th century BCE, when they were inventing the beginning of modern mathematics with theorems about geometry.

We are sometimes critical of the girls' discussions because they do not say exactly what we would like them to say about the topics we present to them. They do not appreciate the subtle relationships to the same extent we think we would in their place. However, they are just starting out what could become a lifetime of interest in math and science. The point is, that they are *talking about math*. That is something that is highly unusual in our day, especially among that age cohort.

Before the VMT Project, we had designed and explored software environments for supporting discussion in small groups and classes. The dominant form of support for group discussion is asynchronous, as in discussion forums, email exchanges, BlackBoard, Knowledge Forum, etc. Even today, if for instance organizers of MOOCs want to add a social aspect to what is generally a talking-head lecture presentation, they turn to asynchronous exchanges. However, especially among today's students, reading and writing prose is avoided, while text chat is considered fun and engaging. Therefore, the VMT Project opted for a synchronous text-chat approach (Stahl, 2009a).

For math discussions, it quickly became apparent that visual representations were necessary. Furthermore, they had to be shared by the members of a discussion. It should be possible for everyone to contribute to drawing the visualization and to pointing to elements of it. Everyone should see the same thing so they can refer to it in their discussion.

Historically, the Math Forum started as the Geometry Forum, a sister project to the creation of Geometer's Sketchpad, both under the direction of Gene Klotz, a professor of mathematics at Swarthmore College. The Math Forum has supported the use of Geometer's Sketchpad for years, so it was natural that we would try to incorporate dynamic geometry into an environment for math talk. We eventually ported an open-source version, GeoGebra, into our online collaboration environment and converted the geometry application into a multi-user system, so a group of users could share the viewing, dragging and construction of figures. This provided a shared object for discussion and rich material for collaborative learning.

Our next focus was on designing topics to guide group discussions. The Math Forum had always had Problems-of-the-Week (PoW) as a core service. Our Virtual Math Teams (VMT) service took this to a new level. Rather than offering independent challenge problems for individual students to solve, we developed sequences of topics for groups to successively explore, leading to a form of curricular coverage.

By arranging for the same groups of students to meet online for a series of topics, we could facilitate progress by the groups in their ability to navigate the software, to work together effectively and to discuss mathematical themes. The hardest thing in our busy lives is to get groups of people to meet at the same time and to focus on a common topic. School classes are one place where this is occasionally possible—although there are many constraints on school time as well.

Given the problems of education today, it is hard to get students to discuss math in a sustainable fashion. However, we have documented that it is possible to achieve that. Here are some lessons from our experience with VMT:

- Synchronous text chat can be a good medium for small-group discussion over the Internet.

- A shared visual representation of discussion material provides an effective focus.

- Carefully designed topics are necessary to guide discussion around curriculum.

- Students can have rich mathematical discussions on their own, with no facilitator present.

- Groups can improve their discussions along many dimensions with guided practice.

- Discussions are most collaborative in groups of 3 or 4; forming groups of about 5 increases the chances that 3 or 4 will always be present and active in a series of meetings.

- Group members tend to play various roles in discussions. These include posing questions, proposing ideas, responding, introducing math facts, constructing figures, symbolizing, directing focus, keeping schedule, reflecting on the discussion, encouraging participation. In groups larger than dyads, the roles can shift among participants, improving individual skills as well as group practices.

- With good scaffolding, groups can construct their own meanings and understandings of topics in mathematics, aligned with accepted views.

- By working on topics as a group, taking turns step by step, sharing each step and building on each other, a group can accomplish more sustained and complicated tasks than any one of the participants would have.

While there is always room for improvement, it seems that success in getting students to talk about math is within reach. We know the basics of how to do this and we have demonstrated its possibility and practicality.

In particular, we have developed a pedagogical model that promises to be scalable. Each year, the Math Forum and associated schools of education offer a teacher professional development course to practicing math teachers. The teachers in this course form into small online teams and engage in sessions using the VMT topics, much like their students will. In addition, the teachers discuss relevant research papers about collaborative learning and dynamic geometry education. They also download logs of their group work and share postings about these. The following semester, the teachers can form groups of their students into virtual math teams to proceed through the VMT curriculum in a series of sessions. The student teams might meet during math class, in an after-school math club or communicate from home computers. The teachers motivate and organize the student groups, as well as providing feedback to them, primarily outside the online sessions. The VMT curriculum is designed to make sense before, while or after the students take a traditional geometry course in school. The teachers can coordinate the sessions with individual practice in GeoGebra and with teacher-directed classroom activities. This suggests a scalable alternative to most MOOCs. It provides research-based content in a context that does not require local teacher involvement. However, it also supports this with trained teacher facilitation outside of the online sessions. In addition, it solves the group formation and individual assessment issues of MOOCs by involving local teachers for those aspects. Above all, it situates learning in a social, collaborative setting—in contrast to the isolating approach of traditional MOOCs. The analysis in this book has documented the potential of such a pedagogical model.

The review of the Cereal Team's efforts at the detailed granularity of the responses of utterances to each other suggests a number of implications for re-design. The VMT curriculum for collaborative dynamic geometry was first formulated in *Dynamic-Geometry Activities with GeoGebra for Virtual Math Teams* (GerryStahl.net/elibrary/topics/activities.pdf) for trial in WinterFest 2012. The Cereal Team used the curriculum defined in *Topics in Dynamic Geometry for Virtual Math Teams* (GerryStahl.net/elibrary/topics/topics.pdf) for WinterFest 2013. This was revised based on general impressions (expanded in this book) of the Cereal Team and the other groups in their cohort for the curriculum in *Explore Dynamic Geometry Together* (GerryStahl.net/elibrary/topics/explore.pdf) for WinterFest 2014. Based on that analysis and suggestions contained in this book, the curriculum has been further refined in *Construct Dynamic Geometry Together* (GerryStahl.net/elibrary/topics/construct.pdf) for WinterFest 2015. In each WinterFest, approximately a hundred students participate in online groups organized by teachers who have taken the corresponding teacher professional development course. The cycles of re-design, trial and analysis are continuing.

The curriculum in *Construct Dynamic Geometry Together* has recently been put into a game format in a GeoGebraBook: *The Construction Crew Game* (http://ggbtu.be/b154045). Unfortunately, the GeoGebraBook is not yet a collaborative medium, like VMT. However, the 50 GeoGebra files that make up the book are available from GeoGebraTube and can be opened in VMT chat rooms for collaboration, including in a new mobile-VMT version for iPads and tablets. Alternatively, small groups could sit around a single display (iPad, laptop, smart board or tabletop display) of the GeoGebraBook and work through the topics by sharing mouse control. Hopefully, GeoGebra will soon release a multi-user version based on VMT.

This section has listed some practical implications for future math education from the VMT Project. However, the major lessons of the project are contained in the lists of group practices enumerated in the core chapters of this book. Groups of students like the Cereal Team can be guided within an online mathematical knowledge-building environment like Virtual Math Teams to develop their mathematical group cognition. This development can be observed to take place primarily through the adoption of specific group practices, such as those witnessed in the preceding chapters: practices of collaboration, dragging, construction, tool usage, dependencies and mathematical discourse or action.

Through the analysis in this book, we have seen how a particular virtual math team learned about dynamic geometry collaboratively. We have identified their enactment of a number of group practices for collaboration and for mathematics. We have also seen in some detail the development of an understanding of mathematical dependency relationships by the students. This understanding has not been as robust as we might want. However, the analysis not only shows the strengths and weaknesses of their understanding, but also suggests the kinds of additional team activities and group practices that might be effective in helping students to deepen their grasp of dependency, given changes in the curriculum based on the analysis.

By supporting student teams to adopt specific group practices, the development of mathematical group cognition can be promoted. One insightful example of this has been provided by the longitudinal analysis of the Cereal Team constructing dynamic triangles together.

Acknowledgments

The data analyzed here is from WinterFest 2013, an effort of the Virtual Math Teams Project, a long-term collaboration among researchers at Drexel University, the Math Forum and Rutgers University at Newark. WinterFest 2013 involved teachers in New Jersey who were participating in a professional development course at Rutgers, and their students.

The analysis of the data was discussed at many weekly research meetings of the VMT project team. Regular participants in the VMT discussions were: Stephen Weimar, Annie Fetter, Tony Mantoan, Michael Khoo, Sean Goggins, Diler Öner, Murat Çakir, Arthur Powell, Muteb Alqahtani and Loretta Dicker. They each contributed in multiple ways to the generation and analysis of the data. In particular, the analysis of Session 3 herein borrows from the parallel analysis in (Öner & Stahl, 2015a) and the analysis of Session 8 incorporates content and figures from (Çakir & Stahl, 2015). Excerpts of this data were also discussed at an all-day workshop at ICLS 2014 (Stahl, 2014d), which is documented at www.gerrystahl.net/vmt/icls2014, including the full dataset for the Cereal Team.

The VMT Project was generously funded by a sequence of grants from the US National Science Foundation covering 2003-2016.

Figures

Logs

References

Adorno, T. W., & Horkheimer, M. (1945). *The dialectic of enlightenment* (J. Cumming, Trans.). New York, NY: Continuum.

Arzarello, F., Olivero, F., Paola, D., & Robutti, O. (2002). A cognitive analysis of dragging practises in Cabri environments. *International Reviews on Mathematical Education (ZDM). 34*(3), 66 - 72.

Barron, B. (2000). Achieving coordination in collaborative problem-solving groups. *Journal of The Learning Sciences. 9*(4), 403-436.

Boaler, J. (2008). *What's math got to do with it? Helping children learn to love their most hated subject: And why it is important for America.* New York, NY: Viking.

Bourdieu, P. (1972/1995). *Outline of a theory of practice* (R. Nice, Trans.). Cambridge, UK: Cambridge University Press.

Bruner, J. (1990). Entry into meaning. In *Acts of meaning.* (pp. 67-97): Harvard U Press.

Çakir, M. P., & Stahl, G. (2013). The integration of mathematics discourse, graphical reasoning and symbolic expression by a virtual math team. In D. Martinovic, V. Freiman & Z. Karadag (Eds.), *Visual mathematics and cyberlearning.* (pp. 49-96). New York, NY: Springer.

Çakir, M. P., & Stahl, G. (2015). *Dragging as a referential resource for mathematical meaning making in a collaborative dynamic-geometry environment [nominated for best paper of conference].* In the proceedings of the CSCL 2015. Gothenburg, Sweden. Web: http://GerryStahl.net/pub/cscl2015cakir.pdf.

Çakir, M. P., Zemel, A., & Stahl, G. (2009). The joint organization of interaction within a multimodal CSCL medium. *International Journal of Computer-Supported Collaborative Learning. 4*(2), 115-149.

Carreira, S., Jones, K., Amado, N., Jacinto, H., & Nobre, S. (2016). *Youngsters solving mathematical problems with technology: The results and implications of the problem@Web project.* New York: NY: Springer.

CCSSI. (2011). High school -- geometry. In Common Core State Standards Initiative (Ed.), *Common core state standards for mathematics.* (pp. 74-78)

Charles, E. S., & Shumar, W. (2009). Student and team agency in VMT. In G. Stahl (Ed.), *Studying virtual math teams.* (ch. 11, pp. 207-224). New York, NY: Springer.

Clark, H., & Brennan, S. (1991). Grounding in communication. In L. Resnick, J. Levine & S. Teasley (Eds.), *Perspectives on socially-shared cognition.* (pp. 127-149). Washington, DC: APA.

Cobb, P. (1995). Mathematical learning and small-group interaction: Four case studies. In P. Cobb & H. Bauersfeld (Eds.), *The emergence of mathematical meaning.* (pp. 25-130). Mahwah, NJ: Lawrence Erlbaum Associates.

Confrey, J., Maloney, A., Nguyen, K., Mojica, G., & Myers, M. (2009). Equipartitioning/splitting as a foundation of rational number reasoning using learning trajectories. Presented at the International Group for the Psychology of Mathematics Education, Thessaloniki, Greece. *Proceedings* pp. 345-352.

Damsa, C. I. (2014). The multi-layered nature of small-group learning: Productive interactions in object-oriented collaboration. *International Journal of Computer-Supported Collaborative Learning. 9*(3), 247-281.

DBR Collective. (2003). Design-based research: An emerging paradigm for educational inquiry. *Educational Researcher. 32*(1), 5-8.

Descartes, R. (1633). *Discourse on method and meditations on first philosophy.* New York, NY: Hackett.

deVilliers, M. (2003). *Rethinking proof with the Geometer's Sketchpad.* Emeryville, CA: Key Curriculum Press.

deVilliers, M. (2004). Using dynamic geometry to expand mathematics teachers' understanding of proof. *International Journal of Mathematics Education in Science & Technology. 35*(4), 703-724.

Dillenbourg, P., Baker, M., Blaye, A., & O'Malley, C. (1996). The evolution of research on collaborative learning. In P. Reimann & H. Spada (Eds.), *Learning in humans and machines: Towards an interdisciplinary learning science.* (pp. 189-211). Oxford, UK: Elsevier.

Donald, M. (2001). *A mind so rare: The evolution of human consciousness.* New York, NY: W. W. Norton.

Dreyfus, H. (1992). *What computers still can't do: A critique of artificial reason.* Cambridge, MA: MIT Press.

Emirbayer, M., & Mische, A. (1998). What is agency? *American Journal of Sociology. 103*(4), 962-1023.

Engeström, Y. (1999). Activity theory and individual and social transformation. In Y. Engeström, R. Miettinen & R.-L. Punamäki (Eds.), *Perspectives on activity theory.* (pp. 19-38). Cambridge, UK: Cambridge University Press.

Engeström, Y. (2008). *From teams to knots.* Cambridge, UK: Cambridge University Press.

Euclid. (300 BCE). *Euclid's elements* (T. L. Heath, Trans.). Santa Fe, NM: Green Lion Press.

Fischer, F., Mandl, H., Haake, J., & Kollar, I. (Eds.). (2006). *Scripting computer-supported collaborative learning: Cognitive, computational and educational perspectives.* Dordrecht, Netherlands: Kluwer Academic Publishers. Computer-supported collaborative learning book series, vol 6.

Gadamer, H.-G. (1960/1988). *Truth and method.* New York, NY: Crossroads.

Gallese, V., & Lakoff, G. (2005). The brain's concepts: The role of the sensory-motor system in conceptual knowledge. *Cognitive Neuropsychology. 21*(3-4), 455-479.

Garfinkel, H. (1967). *Studies in ethnomethodology.* Englewood Cliffs, NJ: Prentice-Hall.

Garfinkel, H. (2002). *Ethnomethodology's program: Working out durkeim's aphorism.* Lanham, Md.: Rowman & Littlefield Publishers.

Gibson, J. J. (1979). *The ecological approach to visual perception.* Boston, MA: Houghton Mifflin.

Goldenberg, E. P., & Cuoco, A. A. (1998). What is dynamic geometry? In R. Lehrer & D. Chazan (Eds.), *Designing learning environments for developing understanding of geometry and space.* (pp. 351-368). Mahwah, NJ: Lawrence Erlbaum.

Goodwin, C. (1994). Professional vision. *American Anthropologist. 96*(3), 606-633.

Goodwin, C., & Heritage, J. (1990). Conversation analysis. *Annual Review of Anthropology. 19*, 283-307.

Hanks, W. (1992). The indexical ground of deictic reference. In A. Duranti & C. Goodwin (Eds.), *Rethinking context: Language as an interactive phenomenon.* (pp. 43-76). Cambridge, UK: Cambridge University Press.

Heath, T. (1921). *A history of Greek mathematics* (Vol. I: From Thales to Euclid). Oxford, UK: Clarendon Press.

Heidegger, M. (1927). *Being and time: A translation of Sein und Zeit* (J. Stambaugh, Trans.). Albany, NY: SUNY Press.

Hölzl, R. (1996). How does "dragging" affect the learning of geometry. *International Journal of Computers for Mathematical Learning. 1*(2), 169–187.

Hoyles, C., & Jones, K. (1998). Proof in dynamic geometry contexts. In C. M. a. V. Villani (Ed.), *Perspectives on the teaching of geometry for the 21st century.* (pp. 121-128). Dordrecht: Kluwer.

Hoyles, C., & Noss, R. (1992). A pedagogy for mathematical microworlds. *Educational Studies in Mathematics. 23*(1), 31-57.

Husserl, E. (1936/1989). The origin of geometry (D. Carr, Trans.). In J.
    Derrida (Ed.), *Edmund Husserl's origin of geometry: An introduction.*
    (pp. 157-180). Lincoln, NE: University of Nebraska Press.

Jones, K. (1997). *Children learning to specify geometrical relationships using
    a dynamic geometry package.* In the proceedings of the 21st
    Conference of the International Group for the Psychology of
    Mathematics Education. P. E. University of Helsinki, Finland.
    Proceedings pp. 3: 121-128.

Jones, K. (2000). Providing a foundation for deductive reasoning: Students'
    interpretations when using dynamic geometry software and their
    evolving mathematical explanations. *Educational Studies in
    Mathematics. 44*(1/2), 55-85.

Jordan, B., & Henderson, A. (1995). Interaction analysis: Foundations and
    practice. *Journal of the Learning Sciences. 4*(1), 39-103.

Khoo, M., & Stahl, G. (2015). *Constructing knowledge: A community of
    practice approach to evaluation in the VMT project.* In the
    proceedings of the CSCL 2015. Gothenburg, Sweden. Web:
    http://GerryStahl.net/pub/cscl2015khoo.pdf.

King, J., & Schattschneider, D. (1997). Making geometry dynamic. In J. King
    & D. Schattschneider (Eds.), *Geometry turned on.* (pp. ix-xiv).
    Washington, DC: Mathematical Association of America.

Kobbe, L., Weinberger, A., Dillenbourg, P., Harrer, A., Hamalainen, R.,
    Hakkinen, P., et al. (2007). Specifying computer-supported
    collaboration scripts. *International Journal of Computer-Supported
    Collaborative Learning. 2*(2-3), 211-224.

Koschmann, T., Kuutti, K., & Hickman, L. (1998). The concept of breakdown
    in Heidegger, leont'ev, and dewey and its implications for education.
    Mind, Culture, and Activity. 5(1), 25-41.

Koschmann, T., Stahl, G., & Zemel, A. (2007). The video analyst's manifesto
    (or the implications of Garfinkel's policies for the development of a
    program of video analytic research within the learning sciences). In
    R. Goldman, R. Pea, B. Barron & S. Derry (Eds.), *Video research in
    the learning sciences.* (pp. 133-144). Mahway, NJ: Lawrence
    Erlbaum Associates.

Koschmann, T., & Zemel, A. (2006). *Optical pulsars and black arrows:
    Discovery's work in 'hot' and 'cold' science.* In the proceedings of the
    International Conference of the Learning Sciences (ICLS 2006).
    Bloomington, IN. Proceedings pp. 356-362.

Laborde, C. (2004). The hidden role of diagrams in pupils' construction of
    meaning in geometry. In C. H. J. Kilpatrick, & O. Skovsmose (Ed.),
    Meaning in mathematics education. (pp. 1-21). Dordrecht,
    Netherlands: Kluwer Academic Publishers.

Lakoff, G. (1987). *Women, fire, and dangerous things*. Chicago: University of Chicago Press.

Lakoff, G., & Núñez, R. (2000). *Where mathematics comes from: How the embodied mind brings mathematics into being*. New York City, NY: Basic Books.

Latour, B. (2007). *Reassembling the social: An introduction to actor-network-theory*. Cambridge, UK: Cambridge University Press.

Latour, B. (2008). The Netz-works of Greek deductions. *Social Studies of Science. 38*(3), 441-459.

Latour, B. (2013). *An inquiry into modes of existence: An anthropology of the modern* (C. Porter, Trans.). Cambridge, MA: Harvard University Press.

Lave, J., & Wenger, E. (1991). *Situated learning: Legitimate peripheral participation*. Cambridge, UK: Cambridge University Press.

LeBaron, C. (2002). Technology does not exist independent of its use. In T. Koschmann, R. Hall & N. Miyake (Eds.), *CSCL 2: Carrying forward the conversation.* (pp. 433-439). Mahwah, NJ: Lawrence Erlbaum Associates.

Lehrer, R., & Schauble, L. (2012). Seeding evolutionary thinking by engaging children in modeling its foundations. *Science Education. 96*, 701-724.

Lemke, J. L. (1993). *Talking science: Language, learning and values*. Norwood, NJ: Ablex.

Lerner, G. (1993). Collectivities in action: Establishing the relevance of conjoined participation in conversation. *Text. 13*(2), 213-245.

List, C., & Pettit, P. (2011). *Group agency: The possibility, design, and status of corporate agents*. Oxford, UK: Oxford University Press.

Livingston, E. (1986). *The ethnomethodological foundations of mathematics*. London, UK: Routledge & Kegan Paul.

Lockhart, P. (2009). *A mathematician's lament: How school cheats us out of our most fascinating and imaginative art forms*. New York, NY: Belevue Literary Press.

Lonchamp, J. (2012). An instrumental perspective on CSCL systems. *International Journal of Computer-Supported Collaborative Learning. 7*(2), 211-237.

Looi, C. K., So, H. J., Toh, Y., & Chen, W. L. (2011). The Singapore experience: Synergy of national policy, classroom practice and design research. *International Journal of Computer-Supported Collaborative Learning. 6*(1), 9-37.

Maxwell, J. (2004). Causal explanation, qualitative research, and scientific inquiry in education. *Educational Researcher. 33*(2), 3-11.

Merleau-Ponty, M. (1945/2002). *The phenomenology of perception* (C. Smith, Trans. 2 ed.). New York, NY: Routledge.

Netz, R. (1999). *The shaping of deduction in Greek mathematics: A study in cognitive history.* Cambridge, UK: Cambridge University Press.

Öner, D. (2008). Supporting students' participation in authentic proof activities in computer supported collaborative learning (CSCL) environments. *International Journal of Computer-Supported Collaborative Learning. 3*(3), 343-359.

Öner, D., & Stahl, G. (2015a). Poster: Tracing the change in discourse in a collaborative dynamic-geometry environment: From visual to more mathematical. Presented at the CSCL 2015, Gothenburg, Sweden. Web: http://GerryStahl.net/pub/cscl2015oner.pdf.

Öner, D., & Stahl, G. (2015b). Tracing the change in discourse in a collaborative dynamic-geometry environment: From visual to more mathematical. *Educational Studies in Mathematics.*

Ong, W. (1998). *Orality and literacy: The technologizing of the world.* New York, NY: Routledge.

Orr, J. (1990). Sharing knowledge, celebrating identity: War stories and community memory in a service culture. In D. S. Middleton & D. Edwards (Eds.), *Collective remembering: Memory in society.* Beverly Hills, CA: SAGE Publications.

Overdijk, M., van Diggelen, W., Andriessen, J., & Kirschner, P. A. (2014). How to bring a technical artifact into use: A micro-developmental perspective. *International Journal of Computer-Supported Collaborative Learning. 9*(3), 283-303.

Phillips, D. C. (2014). Research in the hard sciences, and in very hard "softer" domains. *Educational Researcher. 43*(1), 9-11.

Plato. (340 BCE). *The republic* (F. Cornford, Trans.). London, UK: Oxford University Press.

Plato. (350 BCE). Meno. In E. Hamilton & H. Cairns (Eds.), *The collected dialogues of Plato.* (pp. 353-384). Princeton, NJ: Princeton University Press.

Polanyi, M. (1966). *The tacit dimension.* Garden City, NY: Doubleday.

Polya, G. (1945/1973). *How to solve it: A new aspect of mathematical method.* Princeton, NJ: Princeton University Press.

Rabardel, P., & Beguin, P. (2005). Instrument mediated activity: From subject development to anthropocentric design. *Theoretical Issues in Ergonomics Science. 6*(5), 429–461429–461461.

Rabardel, P., & Bourmaud, G. (2003). From computer to instrument system: A developmental perspective. *Interacting with Computers. 15*, 665–691.

Reckwitz, A. (2002). Toward a theory of social practices : A development in culturalist theorizing. *European Journal of Social Theory. 5*, 243–263.

Renninger, K. A., & Shumar, W. (2002). Community building with and for teachers at the math forum. In K. A. Renninger & W. Shumar (Eds.), *Building virtual communities.* (pp. 60-95). Cambridge, UK: Cambridge University Press.

Renninger, K. A., & Shumar, W. (2004). The centrality of culture and community to participant learning at and with the math forum. In S. Barab, R. Kling & J. H. Gray (Eds.), *Designing for virtual communities in the service of learning.* Cambridge, UK: Cambridge University Press.

Ritella, G., & Hakkarainen, K. (2012). Instrumental genesis in technology-mediated learning: From double stimulation to expansive knowledge practices. *International Journal of Computer-Supported Collaborative Learning. 7*(2), 239-258.

Rosch, E. H. (1973). Natural categories. *Cognitive Psychology. 4*, 328-350.

Roschelle, J., & Teasley, S. (1995). The construction of shared knowledge in collaborative problem solving. In C. O'Malley (Ed.), *Computer-supported collaborative learning.* (pp. 69-197). Berlin, Germany: Springer Verlag.

Roth, W.-M. (2003). *Towards an anthropology of graphing: Semiotic and activity-theoretic perspectives.* The Netherlands: Kluwer Academic Publishers.

Rousseau, J.-J. (1762). *Of the social contract, or principles of political right (du contrat social ou principes du droit politique)* Amsterdam: Marc Michael Rey.

Sacks, H. (1965). *Lectures on conversation.* Oxford, UK: Blackwell.

Sacks, H. (1992). *Lectures on conversation.* Oxford, UK: Blackwell.

Sacks, H., Schegloff, E. A., & Jefferson, G. (1974). A simplest systematics for the organization of turn-taking for conversation. *Language. 50*(4), 696-735.

Sarmiento, J., & Stahl, G. (2008a). *Extending the joint problem space: Time and sequence as essential features of knowledge building [nominated for best paper of the conference].* In the proceedings of the International Conference of the Learning Sciences (ICLS 2008). Utrecht, Netherlands. Web: http://GerryStahl.net/pub/icls2008johann.pdf.

Sarmiento, J., & Stahl, G. (2008b). Group creativity in inter-action: Referencing, remembering and bridging. *International Journal of Human-Computer Interaction (IJHCI).* 492–504.

Sawyer, R. K. (Ed.). (2014). *Cambridge handbook of the learning sciences.* (2nd ed.). Cambridge, UK: Cambridge University Press.

Scardamalia, M. (2002). Collective cognitive responsibility for the advancement of knowledge. In B. Smith (Ed.), *Liberal education in a knowledge society*. Chicago, IL: Open Court.

Scardamalia, M., & Bereiter, C. (2014). Knowledge building and knowledge creation: Theory, pedagogy and technology. In K. Sawyer (Ed.), *Cambridge handbook of the learning sciences*. (2nd ed.). Cambridge, UK: Cambridge University Press.

Schatzki, T. R., Knorr-Cetina, K., & Savigny, E. v. (Eds.). (2001). *The practice turn in contemporary theory*. New York, NY: Routledge.

Schegloff, E., & Sacks, H. (1973). Opening up closings. *Semiotica. 8*, 289-327.

Schegloff, E. A. (1990). On the organization of sequences as a source of 'coherence' in talk-in-interaction. In B. Dorval (Ed.), *Conversational organization and its development*. (pp. 51-77). Norwood, NJ: Ablex.

Schegloff, E. A. (2007). *Sequence organization in interaction: A primer in conversation analysis*. Cambridge, UK: Cambridge University Press.

Scher, D. (2002). *Students' conceptions of geometry in a dynamic geometry software environment*. Unpublished Dissertation, Ph.D., School of Education, New York University. New York, NY. Web: http://GerryStahl.net/pub/GSP_Scher_Dissertation.pdf.

Schmidt, K., & Bannon, L. (1992). Taking CSCW seriously: Supporting articulation work. *CSCW. 1*(1), 7-40.

Schön, D. A. (1983). *The reflective practitioner: How professionals think in action*. New York, NY: Basic Books.

Seddon, C. (2014). *Humans: From the beginning: From the first apes to the first cities*. Kindle, ebook: Glanville Publications.

Sfard, A. (1994). Reification as the birth of metaphor. *For the Learning of Mathematics. 14*(1), 44-55.

Sfard, A. (2002). There is more to discourse than meets the ears: Looking at thinking as communicating to learn more about mathematical learning. In C. Kieran, E. Forman & A. Sfard (Eds.), *Learning discourse: Discursive approaches to research in mathematics education*. (pp. 13-57). Dordrecht, Netherlands: Kluwer.

Sfard, A. (2008a). *Learning mathematics as developing a discourse*. In the proceedings of the ICME 11. Monterrey, Mexico.

Sfard, A. (2008b). *Thinking as communicating: Human development, the growth of discourses and mathematizing*. Cambridge, UK: Cambridge University Press.

Sfard, A., & Cobb, P. (2014). Research in mathematics education: What can it teach us about human learning? In K. Sawyer (Ed.), *Cambridge handbook of the learning sciences*. (2nd ed., pp. 545-564). Cambridge, UK: Cambridge University Press.

Sfard, A., & Kieran, C. (2001). Cognition as communication: Rethinking learning-by-talking through multi-faceted analysis of students' mathematical interactions. *Mind, Culture, and Activity. 8*(1), 42-76.

Shannon, C., & Weaver, W. (1949). *The mathematical theory of communication*. Chicago, Il: University of Illinois Press.

Sinclair, N. (2008). *The history of the geometry curriculum in the united states*. Charlotte, NC: Information Age Publishing, Inc.

Stahl, G. (1993). *Interpretation in design: The problem of tacit and explicit understanding in computer support of cooperative design.* Unpublished Dissertation, Ph.D., Department of Computer Science, University of Colorado. Boulder, CO. Web: http://GerryStahl.net/publications/dissertations/computer or http://GerryStahl.net/elibrary/tacit

Stahl, G. (2000). *A model of collaborative knowledge-building.* In the proceedings of the Fourth International Conference of the Learning Sciences (ICLS '00). B. Fischman & S. O'Conner-Divelbiss. Ann Arbor, MI. Proceedings pp. 70-77. Lawrence Erlbaum Associates. Web: http://GerryStahl.net/pub/icls2000.pdf.

Stahl, G. (2005). *Group cognition: The collaborative locus of agency in CSCL.* In the proceedings of the international conference on Computer Support for Collaborative Learning (CSCL '05). T. Koschmann, D. Suthers & T.-W. Chan. Taipei, Taiwan. Proceedings pp. 632-640. Lawrence Erlbaum Associates. Web: http://GerryStahl.net/pub/cscl2005.pdf & http://GerryStahl.net/pub/cscl2005ppt.pdf.

Stahl, G. (2006). *Group cognition: Computer support for building collaborative knowledge.* Cambridge, MA: MIT Press.

Stahl, G. (2009a). A chat about chat. In G. Stahl (Ed.), *Studying virtual math teams.* (ch. 1, pp. 7-16). New York, NY: Springer.

Stahl, G. (2009b). *Studying virtual math teams.* New York, NY: Springer.

Stahl, G. (2011a). *How a virtual math team structured its problem solving.* In the proceedings of the Connecting computer-supported collaborative learning to policy and practice: CSCL 2011 conference proceedings. N. Miyake, H. Spada & G. Stahl. Lulu: ISLS. Proceedings pp. 256-263. Web: http://GerryStahl.net/pub/cscl2011stahl.pdf, http://GerryStahl.net/pub/cscl2011stahl.ppt.pdf, http://youtu.be/0Dg02YQCQIE.

Stahl, G. (2011b). How I view learning and thinking in CSCL groups. *Research and Practice in Technology Enhanced Learning (RPTEL). 6*(3), 137-159.

Stahl, G. (2011c). Social practices of group cognition in virtual math teams. In S. Ludvigsen, A. Lund, I. Rasmussen & R. Säljö (Eds.), *Learning* across sites: New tools, infrastructures and practices. (ch. 12, pp. 190-205). New York, NY: Routledge.

Stahl, G. (2011d). *The structure of collaborative problem solving in a virtual math team.* In the proceedings of the iConference 2011. Seattle, WA. Web: http://GerryStahl.net/pub/iconf2011.pdf.

Stahl, G. (2011e). *Theories of cognition in CSCW.* In the proceedings of the European Computer-Supported Cooperative Work. Aarhus, Denmark. Web: http://GerryStahl.net/pub/ecscw2011.pdf.

Stahl, G. (2012a). *Dynamic-geometry activities with GeoGebra for virtual math teams.* Web: http://GerryStahl.net/elibrary/topics/activities.pdf.

Stahl, G. (2012b). Ethnomethodologically informed. *International Journal of Computer-Supported Collaborative Learning. 7*(1), 1-10.

Stahl, G. (2013a). Seminar: Analyzing virtual math teams enacting geometric practices. Presented at the LinCS Seminars, University of Gothenburg, Sweden. Web: http://GerryStahl.net/pub/analyzing.pdf.

Stahl, G. (2013b). *Topics in dynamic geometry for virtual math teams.* Web: http://GerryStahl.net/elibrary/topics/topics.pdf.

Stahl, G. (2013c). *Translating Euclid: Designing a human-centered mathematics.* San Rafael, CA: Morgan & Claypool Publishers.

Stahl, G. (2014a). *Construct dynamic geometry together.* Web: http://GerryStahl.net/elibrary/topics/construct.pdf; http://ggbtu.be/b140867.

Stahl, G. (2014b). *The display of learning in groupwork.* In the proceedings of the ACM Conference on Supporting Groupwork (GROUP 2014). Sanibel Island, FL. Web: http://GerryStahl.net/pub/group2014.pdf.

Stahl, G. (2014c). *Explore dynamic geometry together.* Web: http://GerryStahl.net/elibrary/topics/explore.pdf.

Stahl, G. (2014d). Workshop: Interaction analysis of student teams enacting the practices of collaborative dynamic geometry. Presented at the International Conference of the Learning Sciences (ICLS 2014), Boulder, CO. Web: http://GerryStahl.net/pub/icls2014workshop.pdf.

Stahl, G. (2015). *The construction crew game.* Web: http://GerryStahl.net/elibrary/topics/game.pdf; http://ggbtu.be/b154045.

Stahl, G. (2016a). From intersubjectivity to group cognition. *Computer Supported Cooperative Work (CSCW). 25*(4), 355-384.

Stahl, G. (2016b). The group as paradigmatic unit of analysis: The contested relationship of CSCL to the learning sciences. In M. A. Evans, M. J. Packer & R. K. Sawyer (Eds.), *Reflections on the learning sciences.* (ch. 5, pp. 76-102). New York, NY: Cambridge University Press.

Stahl, G., Koschmann, T., & Suthers, D. (2006). Computer-supported collaborative learning: An historical perspective. In R. K. Sawyer

(Ed.), *Cambridge handbook of the learning sciences.* (pp. 409-426). Cambridge, UK: Cambridge University Press.

Stahl, G., Koschmann, T., & Suthers, D. (2014). Computer-supported collaborative learning. In R. K. Sawyer (Ed.), *Cambridge handbook of the learning sciences, revised version.* (ch. 24, pp. 479-500). Cambridge, UK: Cambridge University Press.

Stahl, G., Zhou, N., Çakir, M. P., & Sarmiento-Klapper, J. W. (2011). *Seeing what we mean: Co-experiencing a shared virtual world.* In the proceedings of the Connecting computer-supported collaborative learning to policy and practice: CSCL 2011 conference proceedings. Lulu: ISLS. Proceedings pp. 534-541. Web: http://GerryStahl.net/pub/cscl2011.pdf, http://GerryStahl.net/pub/cscl2011.ppt.pdf, http://youtu.be/HC6eLNNIvCk.

Suchman, L. A. (2007). *Human-machine reconfigurations: Plans and situated actions* (2nd ed.). Cambridge, UK: Cambridge University Press.

Suchman, L. A., & Jordan, B. (1990). Interactional troubles in face-to-face survey interviews. *Journal of the American Statistical Association. 85*, 232-244.

Teasley, S. D., & Roschelle, J. (1993). Constructing a joint problem space: The computer as a tool for sharing knowledge. In S. P. Lajoie & S. J. Derry (Eds.), *Computers as cognitive tools.* (pp. 229-258). Mahwah, NJ: Lawrence Erlbaum Associates, Inc.

Tee, M. Y., & Karney, D. (2010). Sharing and cultivating tacit knowledge in an online learning environment. *International Journal of Computer-Supported Collaborative Learning. 5*(4), 385-413.

Thorndike, E. L. (1914). *Educational psychology* (Vol. I-III). New York, NY: Teachers College.

Tomasello, M. (2014). *A natural history of human thinking.* Cambridge: MA: Harvard University Press.

Turner, S. (1994). *The social theory of practices: Tradition, tacit knowledge, and presuppositions.* Chicago, IL: University of Chicago Press.

van Hiele, P. (1986). *Structure and insight: A theory of mathematics education.* Orlando, FL: Academic Press.

van Hiele, P. (1999). Developing geometric thinking through activities that begin with play. *Teaching Children Mathematics. 310-316.*

Vygotsky, L. (1930). *Mind in society.* Cambridge, MA: Harvard University Press.

Vygotsky, L. (1934/1986). *Thought and language.* Cambridge, MA: MIT Press.

Wee, J. D., & Looi, C.-K. (2009). A model for analyzing math knowledge building in VMT. In G. Stahl (Ed.), *Studying virtual math teams.* (ch. 25, pp. 475-497). New York, NY: Springer.

Wegerif, R. (2007). *Dialogic, education and technology: Expanding the space of learning.* New York, NY: Kluwer-Springer.

Weick, K. E. (1988). Enacted sensemaking in crisis situations. *Journal of Management Studies. 25*(4), 305-317.

Wittgenstein, L. (1953). *Philosophical investigations.* New York, NY: Macmillan.

Yin, R. K. (2004). Case study methods. In *Complementary methods for research in education.* (3rd ed.). Washington, DC: American Educational Research Association.

Yin, R. K. (2009). *Case study research. Design and methods* (4th ed.). Thousand Oaks, CA: Sage Publications.

Zemel, A., & Çakir, M. P. (2009). Reading's work in VMT. In G. Stahl (Ed.), *Studying virtual math teams.* (ch. 14, pp. 261-276). New York, NY: Springer.

Zemel, A., Çakir, M. P., Stahl, G., & Zhou, N. (2009). *Learning as a practical achievement: An interactional perspective.* In the proceedings of the international conference on Computer Support for Collaborative Learning (CSCL 2009). Rhodes, Greece. Web: http://GerryStahl.net/pub/cscl2009zhou.pdf.

Zemel, A., & Koschmann, T. (2013). Recalibrating reference within a dual-space interaction environment. *International Journal of Computer-Supported Collaborative Learning. 8*(1), 65-87.

Zhou, N. (2009). Question co-construction in VMT chats. In G. Stahl (Ed.), *Studying virtual math teams.* (ch. 8, pp. 141-159). New York, NY: Springer.

Zhou, N., Zemel, A., & Stahl, G. (2008). *Questioning and responding in online small groups engaged in collaborative math problem solving.* In the proceedings of the International Conference of the Learning Sciences (ICLS 2008). Utrecht, Netherlands. Web: http://GerryStahl.net/pub/icls2008nan.pdf.

www.ingramcontent.com/pod-product-compliance
Lightning Source LLC
Chambersburg PA
CBHW031829170526
45157CB00001B/231